Cisco® VPN Configuration Gu

Practical Cisco VPN Configuration Tutorials

*Your one-stop Information Resource
For Configuring Cisco VPN Technologies
on Routers and ASA Firewalls*

Written By: Harris Andrea

*MSc Electrical Engineering and Computer Science
Cisco Certified Network Associate (CCNA)
Cisco Certified Network Professional (CCNP)
Cisco Certified Security Professional (CCSP)
Certified Ethical Hacker (CEH)
EC-Council Certified Security Analyst (ECSA)*

http://www.networkstraining.com

Legal Notice:

© 2014, Harris Andrea.
All rights reserved.

Email: **admin@networkstraining.com**
Website: **http://www.networkstraining.com/**

This Book contains material protected under International and Federal Copyright Laws and Treaties. No part of this publication may be transmitted or reproduced in any way without the prior written permission of the author. Violations of this copyright will be enforced to the full extent of the law.

The information services and resources provided in this Book are based upon the current Internet environment as well as the author's experience. The techniques presented here have been proven to be successful. Because technologies are constantly changing, the configurations and examples presented in this Book may change, cease or expand with time. We hope that the skills and knowledge acquired from this Book will provide you with the ability to adapt to inevitable evolution of technological services. However, we cannot be held responsible for changes that may affect the applicability of these techniques. The opinions expressed in this Book belong to the author and are not necessarily those of Cisco Systems, Inc. The author is not affiliated with Cisco Systems, Inc.

All trademarks are trademarks of their respective owners. Rather than putting a trademark symbol after every occurrence of a trademarked name, we use names in an editorial fashion only, and to the benefit of the trademark owner, with no intention of infringement of the trademark. Where such designations appear in this book, they have been printed with initial caps.

All product names, logos and artwork are copyrights of their respective owners. None of the owners have sponsored or endorsed this publication. While all attempts have been made to verify information provided, the author assumes no responsibility for errors, omissions, or contrary interpretation of the subject matter herein. Any perceived slights of peoples or organizations are unintentional. The purchaser or reader of this publication assumes responsibility for the use of these materials and information. No guarantees of income are made. The author reserves the right to make changes and assumes no responsibility or liability whatsoever on behalf of any purchaser or reader of these materials.

ISBN-10: 1-5005-2290-2
ISBN-13: 978-1-5005-2290-2

Table of Contents:

Chapter 1 Introduction to VPN Technologies 8
- 1.1 Policy-Based Vs Route-Based VPN 9
- 1.2 Policy-Based VPN (Traditional IPSEC VPN) 11
  - 1.2.1 What is IPSEC 11
  - 1.2.2 How IPSEC Works 13
  - 1.2.3 Site-to-Site and Hub-and-Spoke IPSEC VPN 13
  - 1.2.4 Remote Access IPSEC VPN 15
- 1.3 Route-Based VPN 16
  - 1.3.1 VPN using GRE 16
    - 1.3.1.1 GRE Vs IPSEC 17
  - 1.3.2 VPN using Virtual Tunnel Interface (VTI) 19
    - 1.3.2.1 Static VTI 20
    - 1.3.2.2 Dynamic VTI 21
- 1.4 Dynamic Multipoint VPN (DMVPN) 23
- 1.5 SSL Based VPNs (WebVPN) 26
  - 1.5.1 Types of SSL Based VPNs 26
  - 1.5.2 Comparison between SSL VPN Technologies 26
  - 1.5.3 Overview of AnyConnect VPN operation: 27
- 1.6 Practical Applications for each VPN Type 29
  - 1.6.1 Policy-Based (Traditional IPSEC) VPN Applications 29
  - 1.6.2 Route-Based GRE VPN Applications 30
  - 1.6.3 Route-Based VTI VPN Applications 31
  - 1.6.4 Dynamic Multipoint VPN Applications 31

Chapter 2 VPN Configuration on Cisco Routers 33
- 2.1 Policy-Based VPN Configuration on Cisco Routers 33
  - 2.1.1 Site-to-Site IPSEC VPN 33
    - 2.1.1.1 Site-to-Site IPSEC VPN with Dynamic IP 42
  - 2.1.2 Hub-and-Spoke IPSEC VPN 44
  - 2.1.3 Remote Access IPSEC VPN 47

2.1.4	Site-to-Site and Remote Access IPSEC VPN on same device	53
2.2	Route-Based VPN Configuration on Cisco Routers	59
2.2.1	Site-to-Site VPN Using GRE with IPSEC Protection	59
2.2.2	Hub-and-Spoke VPN Using GRE with IPSEC Protection	63
2.2.3	VPN Using Static Virtual Tunnel Interface (SVTI)	68
2.2.4	VPN Using Dynamic Virtual Tunnel Interface (DVTI)	69
2.3	Dynamic Multipoint VPN (DMVPN)	76
2.4	PPTP VPN	83

Chapter 3 VPN Configuration on ASA Firewalls 87

3.1	Policy-Based VPN Configuration on Cisco ASA	87
3.1.1	Site-to-Site IPSEC VPN	87
3.1.1.1	Restricting IPSEC VPN Traffic between the Two Sites	94
3.1.2	Hub-and-Spoke IPSEC VPN with Dynamic IP Spoke	96
3.1.2.1	Spoke to Spoke Communication via the Hub ASA	99
3.1.3	IPSEC VPN between Cisco ASA and Cisco Router	102
3.1.4	Remote Access IPSEC VPN	106
3.1.5	Hub-and-Spoke and Remote Access VPN on same device	111
3.1.5.1	Enable Remote Users to Access Spoke Sites through the Hub	115
3.1.6	Site-to-Site IPSEC VPN with failover using backup ISP	117
3.1.7	Site-to-Site IPSEC VPN with Duplicate Subnets –Example1	123
3.1.8	Site-to-Site IPSEC VPN with Duplicate Subnets –Example2	127
3.1.9	Site-to-Site IKEv2 IPSEC VPN	131
3.2	SSL-Based VPN Configuration on Cisco ASA	139
3.2.1	Anyconnect SSL Web VPN	139
3.3	VPN Authentication using External Server	149
3.3.1	VPN Authentication using Microsoft Active Directory	149
3.3.2	VPN Authentication using RADIUS or TACACS	152
3.3.3	VPN Authentication using RSA	154

Chapter 4 Complete Configuration Examples 156

4.1	Complete VPN Configurations on Cisco Routers	156
4.1.1	Site-to-Site IPSEC VPN	156

4.1.2	Site-to-Site IPSEC VPN with Dynamic IP	160
4.1.3	Hub-and-Spoke IPSEC VPN – Static IP Spokes	164
4.1.4	Hub-and-Spoke IPSEC VPN – Dynamic IP Spoke	170
4.1.5	Remote Access IPSEC VPN	173
4.1.6	Site-to-Site and Remote Access IPSEC VPN on same device	176
4.1.7	Site-to-Site VPN using GRE with IPSEC Protection	184
4.1.8	Hub-and-Spoke VPN using GRE with IPSEC Protection	188
4.1.9	Hub-and-Spoke VPN using DVTI and SVTI	195
4.1.10	Dynamic Multipoint VPN (DMVPN)	202
4.1.11	Point to Point Tunnelling Protocol (PPTP)	209
4.2	Complete VPN Configurations on Cisco ASA	211
4.2.1	Site-to-Site IPSEC VPN	211
4.2.2	Hub-and-Spoke IPSEC VPN with Dynamic IP Spoke	216
4.2.3	IPSEC VPN Between Cisco ASA and Cisco Router	223
4.2.4	Remote Access IPSEC VPN on Cisco ASA	228
4.2.5	Hub-and-Spoke and Remote Access VPN on same device	231
4.2.6	Site-to-Site IPSEC VPN with failover using backup ISP	239
4.2.7	Site-to-Site IPSEC VPN with Duplicate Subnets-Example1	245
4.2.8	Site-to-Site IPSEC VPN with Duplicate Subnets-Example2	250
4.2.9	Anyconnect SSL Web VPN	255

About the Author:

Harris Andrea is a Senior Network Security Engineer working for a leading Internet Service Provider in Europe. He graduated from the University of Kansas USA in 1998 with a B.S and M.S degrees in Electrical Engineering and Computer Science. Since then, he has been working in the Networking field, designing, implementing and managing large scale networking projects with Cisco products and technologies. His main focus is on Network Security based on Cisco ASA Firewalls, VPN technologies, IDS/IPS products, AAA services, IOS Security Features etc. To support his knowledge and to build a strong professional standing, Harris pursued and earned several Cisco Certifications such as CCNA, CCNP, CCSP and other security related certifications such as CEH and ECSA. He is also a technology blogger owing a networking blog about Cisco technologies which you can visit for extra technical information and tutorials.

http://www.networkstraining.com

Introduction:

Thank you for purchasing this technical Book about configuring Cisco VPN Technologies. Virtual Private Networks constitute a hot topic in networking because they provide low cost and secure communications while improving productivity by extending corporate networks to remote locations.

The two major Cisco networking devices that support VPNs are Cisco Routers and Cisco ASA Firewalls. That's why this book focuses on VPN implementations using these two device types. I remember building my first site-to-site IPSEC VPN back in 2000 using two Cisco PIX 501 firewalls. I was impressed when communication was established between two private LAN networks over the Internet. Since then, I have designed, configured and managed hundreds of VPN implementations using Cisco Routers and PIX/ASA firewalls. This Book therefore is the result of my working experience with Cisco VPN technology for more than a decade.

I have tried to include the most important and commonly found VPN topologies that you will find in real world networks. Also, I have included several scenarios which are somewhat infrequent or unusual to encounter and they are also a little bit difficult to configure. These include VPN Failover using Backup ISP, site-to-site VPN with duplicate subnets, VPN Hairpinning, Active Directory authentication, DMVPN etc.

Virtual Private Networks are based on complex protocols and algorithms. The intention of this book is not to delve into the theory and details of VPNs but rather to provide practical and step-by-step configuration instructions. Nevertheless, some required basic theory, applications and comparisons of the various VPN types are included in the book. Overall, I believe that this book is probably the most updated and comprehensive resource on Cisco VPNs out there and I firmly believe it will be valuable for Cisco networking professionals.

If you are interested in my other book "**Cisco ASA Firewall Fundamentals-3rd Edition**", you can find more information about it here: http://www.networkstraining.com/ciscoasaebook.php

For any questions that you may have or clarifications about the information presented in this Book, please contact me at: admin@networkstraining.com

Have fun reading my Book. I hope it will be a valuable resource for you.

Chapter 1   Introduction to VPN Technologies

The intention of this book is to be a practical configuration guide of the major VPN technologies supported by Cisco, thus I will not cover all the theory and details behind Virtual Private Networks. However, an introductory description of the various VPN types that we will be using throughout this book is essential. Specifically, I will briefly discuss some theory and practical applications of Policy-Based VPNs (traditional IPSEC VPNs), Route-Based VPNs (GRE VPNs and VPNs based on Virtual Tunnel Interface-VTI), SSL Web VPNs, and finally Dynamic Multipoint VPNs (DMVPN). In the next Chapters we will go into the actual practical configuration details of the various VPN types.

The diagram below illustrates the four general VPN categories that we will be using in this book.

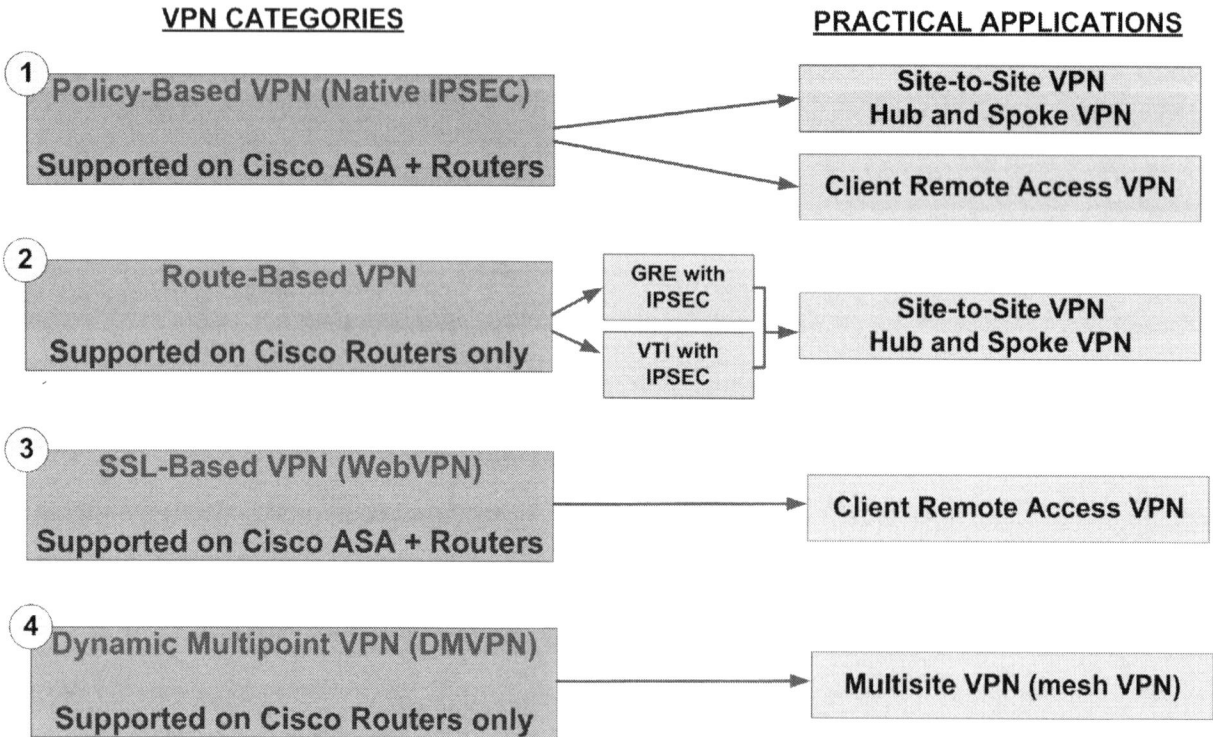

1.1 Policy-Based Vs Route-Based VPN

Two important VPN categories supported by Cisco are the first two shown on figure above. These are **Policy-Based** and **Route-Based** VPNs. In my opinion it's important to describe the main differences between these two VPN types. Knowing the differences will help professionals choose the right VPN type for their company or customers.

Both of these VPN categories make use of the IPSEC protocol (we will describe it later) which is the de facto standard for creating secure VPN networks. Let's see a brief description of them below:

- **Policy-Based IPSEC VPN**: This is the traditional IPSEC VPN type which is still widely used today. This VPN category is supported on both Cisco ASA Firewalls and Cisco Routers. With this VPN type, the device encrypts and encapsulates a subset of traffic flowing through an interface according to a defined policy (using an Access Control List). The IPSEC protocol is used for tunneling and for securing the communication flow. Most of the discussion on IPSEC in this book is based on the legacy IKEv1 IPSEC, although there is a small section about the new IKEv2 IPSEC as well.
- **Route-Based VPN**: A route-based VPN configuration employs Layer3 routed tunnel interfaces as the endpoints of the virtual network. All traffic passing through a special Layer3 tunnel interface is placed into the VPN. Rather than relying on an explicit policy to dictate which traffic enters the VPN, static or dynamic IP routes are configured to direct the desired traffic through the VPN tunnel interface. This configuration method is supported only on Cisco Routers and is based on **GRE** or **VTI** Tunnel Interfaces as we will see later. For secure communication, Route-Based VPNs use also the IPSEC protocol on top of the GRE or VTI tunnel to encrypt everything.

The Table below shows the main differences between **Policy-Based** and **Route-Based** VPNs:

Policy-Based IPSEC VPN (Traditional IPSEC)	Route-Based VPN (GRE and VTI)
Supported on most network devices (Cisco Routers, Cisco ASA, other vendors etc)	Supported only on Cisco IOS Routers. Very Limited interoperability with other vendors.
Does not support multicast or non-IP protocols	Supports multicast (GRE and VTI) and non-IP protocols (GRE)
Routing Protocols (e.g OSPF, EIGRP) <u>cannot</u> pass through the VPN tunnel	Routing Protocols (e.g OSPF, EIGRP) <u>can</u> pass through the VPN tunnel.
Use an access list to select which traffic is going to be encrypted and placed in VPN tunnel.	All traffic passing through a special Tunnel Interface will be encapsulated and placed in the VPN.
Strong Security natively	GRE or VTI alone do not provide security. You must combine them with IPSEC for securing the VPN.
Complex Configuration	Simplified Configuration
Limited QoS.	QoS is fully supported.

1.2 Policy-Based VPN (Traditional IPSEC VPN)

This section discusses Policy-Based VPN using the IPSEC protocol standard. This is the traditional IPSEC VPN used also by many other Vendors in addition to Cisco. IPSEC is supported on both Cisco ASA firewalls (by default) and Cisco Routers (with the proper IOS image).

Traditional IPSEC can be used to build **Site-to-Site** (also called Lan-to-Lan) VPNs and also client **Remote Access** VPNs. The first VPN type (Site-to-Site or Hub-and-Spoke) is used to securely connect together distant LAN networks, while the later (Remote Access VPN) allows remote users/teleworkers to securely communicate with their corporate network.

The legacy IPSEC protocol (IKEv1 IPSEC) has been enhanced with a new IPSEC version, called also IKEv2 IPSEC. In this book we are dealing mostly with the legacy IKEv1 IPSEC because it is still the most widely used all over the world. However, we will briefly describe also the new IKEv2 IPSEC and see a basic configuration scenario with this new type of IPSEC on Cisco ASA firewalls.

1.2.1 What is IPSEC

IP Security (**IPSEc**) is an open IETF standard that enables secure and encrypted communication. It is a suit of protocols that provide data confidentiality, integrity, and authentication. A Virtual Private Network (**VPN**) is a secure private tunnel over an insecure path (e.g over the Internet). IPSEC therefore is ideal to build VPNs over the Internet or over any other non-secure networks. Therefore, you will find IPSEC in most VPN implementations, either used as a tunneling protocol alone (as in Policy-Based VPNs) or in conjunction with GRE or VTI (as in Route-Based VPNs).

IPSEc works at the network layer, encrypting and authenticating IP packets between participating devices (peers), such as Cisco routers, Cisco ASA firewalls, VPN software clients etc. Since IPSEC is an IETF standard, almost all firewall and router vendors support it. Thus, you can use traditional IPSEC to create VPNs between different vendors such as Cisco, Juniper, Checkpoint, Palo Alto, Fortinet, Sonic Wall etc.

NOTE:

One important limitation of traditional IPSEC VPN is that ONLY <u>unicast</u> IP traffic can pass through the VPN tunnel. This means that if you have two or more sites connected over the Internet with IPSEC VPN, you cannot pass multicast or other non-IP protocols (such as IPX or AppleTalk) through the VPN. For example, passing routing protocols (such as OSPF and EIGRP which use multicast) is not possible through an IPSEC tunnel. In order to support multicast traffic you need to use other VPN protocol technologies (such as GRE or VTI using route-based VPN configuration).

The following IPSEc protocols and standards will be used later in our discussion, so it's a good idea to briefly explain their functionality and usage:

- **ESP (Encapsulating Security Payload):** This is the first of the two main protocols that make up the IPSEc standard. It provides data integrity, authentication, and confidentiality services. ESP is used to encrypt the data payload of the IP packets.
- **AH (Authentication Header):** This is the second of the two main protocols of IPSEc. It provides data integrity, authentication, and replay-detection. It does not provide encryption services, but rather it acts as a "digital signature" for the packets to ensure that tampering of data has not occurred.
- **Internet Key Exchange (IKE):** This is the mechanism used by the VPN appliance for securely exchanging encryption keys, authenticating IPSEc peers and negotiating IPSEc Security parameters. On Cisco ASA firewall and Routers, this is synonymous with **ISAKMP** (or **IKEv1**) as we will see in the IPSEc configuration.
- **DES, 3DES, AES:** All these are encryption algorithms supported by Cisco ASA Firewall and Routers. DES is the weakest one (uses 56-bit encryption key), and AES is the strongest one (uses 128, 192, or 256 bit encryption keys). 3DES is a middle choice using 168-bit encryption key.
- **Diffie-Hellman Group (DH):** This is a public-key cryptography protocol used by IKE to establish session keys.
- **MD5, SHA-1:** These are both Hash Algorithms used to authenticate packet data. SHA is stronger than MD5.
- **Security Association (SA):** An SA is a connection between two IPSEc peers. Each IPSEc peer maintains an SA database in its memory containing SA parameters. SAs are uniquely identified by the IPSEc peer address, security protocol, and security parameter index (SPI).

1.2.2 How IPSEC Works

There are five main steps followed by the IPSEc devices:

1. **Interesting Traffic**: The IPSEc devices recognize the traffic to protect using Access Control Lists (in policy-based IPSEC).
2. **Phase 1 (ISAKMP / IKEv1)**: The IPSEc devices negotiate an IKE security policy and establish a secure channel for communication.
3. **Phase 2 (IPSEc)**: The IPSEc devices negotiate an IPSEc security policy to protect data.
4. **Data Transfer**: Data is transferred securely between the IPSEc peers based on the IPSEc parameters and keys negotiated during the previous phases.
5. **IPSEc Tunnel Terminated**: IPSEc SAs terminate when timing out or a certain data volume is reached.

The steps above will become clear when we see actual configuration examples. Let's start with the first IPSEc VPN application that we will describe in this section: Site-to-Site and Hub-and-Spoke IPSEC VPN.

1.2.3 Site-to-Site and Hub-and-Spoke IPSEC VPN

Just for illustration purposes, the diagrams below show a simple site-to-site VPN and a simple Hub-and-Spoke topologies using Cisco ASA firewall devices. In this book we will see how to configure Site-to-Site and Hub-and-Spoke IPSEC VPN topologies using ASA firewalls, Cisco Routers and also combination of Routers with ASA. A Hub-and-Spoke topology is using multiple Site-to-Site VPNs between a central Device (Hub) and remote site devices (Spokes).

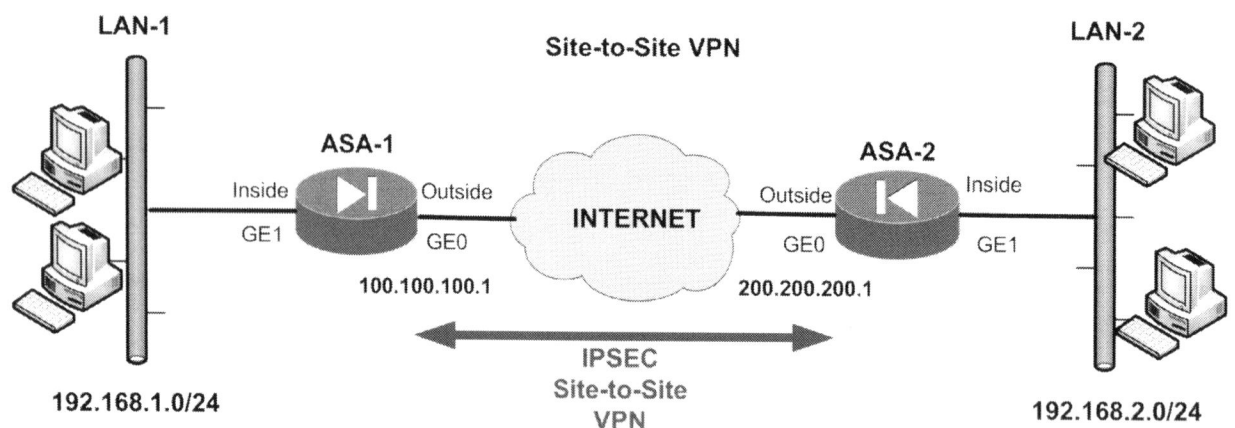

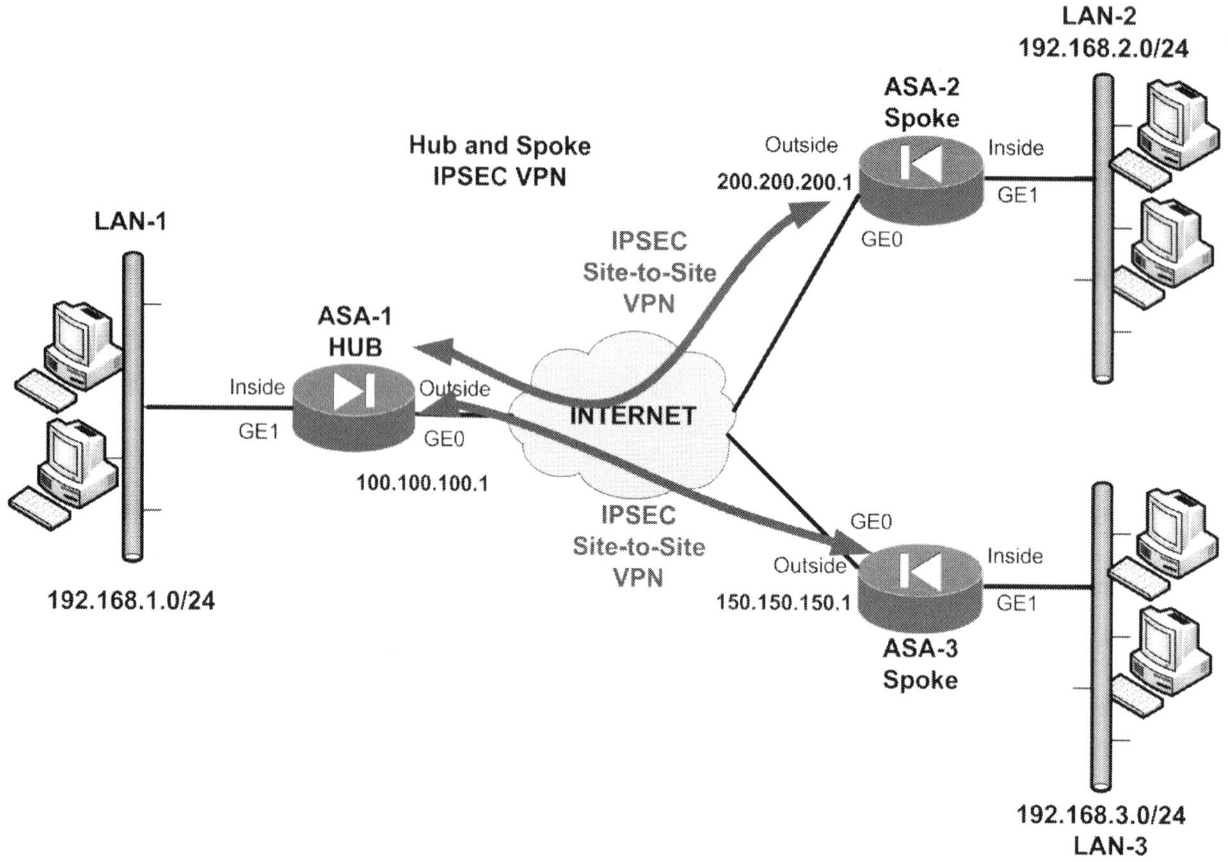

Site-to-Site (and Hub-and-Spoke) IPSEc VPNs are sometimes called LAN-to-LAN VPNs. As the name implies, this VPN type connects together two (or more) distant LAN networks over the Internet. Usually, Local Area Networks use private addressing as shown on our diagram above. Without VPN connectivity, the private LAN networks above (LAN-1, LAN-2, LAN-3) wouldn't be able to communicate. By configuring a Site-to-Site IPSEc VPN between the ASA firewalls, we can establish a secure tunnel over the Internet, and pass our private LAN traffic inside this tunnel. The result is that hosts in network 192.168.1.0/24 can now directly access hosts in 192.168.2.0/24 and in 192.168.3.0/24 networks (and vice-versa) as if they were located in the same LAN. The IPSEc tunnel is established between the Public IP addresses of the firewalls. You can find all configuration details in sections 2.1.1, 2.1.2, 3.1.1, 3.1.2.

1.2.4 Remote Access IPSEC VPN

The second practical application of policy-based IPSEc VPN that we will describe in this section is Remote Access IPSEC VPN using a Cisco VPN client installed on the computer of a remote user. This type of VPN allows remote users/teleworkers with Internet access to establish a secure IPSEc VPN tunnel with their central corporate network. The user must have a Cisco VPN client software installed on his/her computer which will enable a secure communication with the IPSEC enabled device (ASA firewall or Router) in the central office. After the VPN is established between the remote user and the IPSEC-enabled device, the user is assigned a private IP address from a predefined pool, and then gets connected to the Corporate LAN network. All LAN resources can then be accessed remotely. See example diagram below:

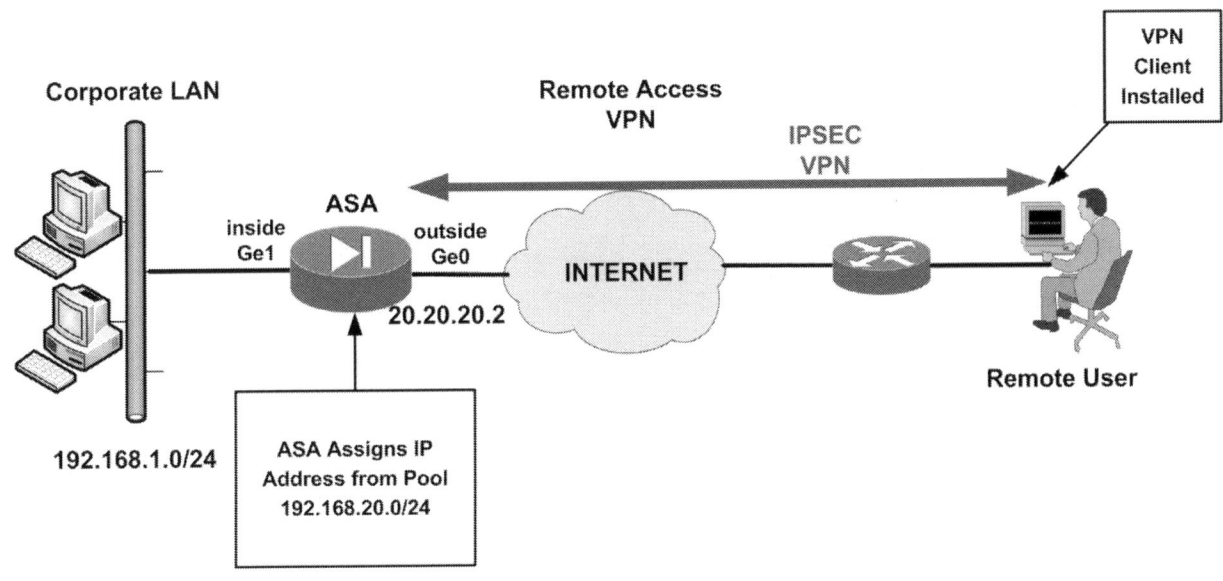

Our example network topology above shows a central ASA firewall (it could be also an IPSEC capable Router) protecting the Corporate LAN, and a remote user with a software VPN client establishing a secure connection with the ASA. An IP address in the range 192.168.20.0/24 will be assigned to the VPN client, which will be allowed to communicate with the Internal Corporate network 192.168.1.0/24. Once the Remote Access VPN is established, the remote user by default will not be able to access anything else on the Internet, except the Corporate LAN network. This behavior can be altered by configuring the "**split tunneling**" feature on the Firewall (or Router), which however is not recommended for security purposes. You can find all configuration details in Sections 2.1.3, 3.1.4.

1.3   Route-Based VPN

Route-Based VPNs are supported only on Cisco routers. A Layer3 virtual Tunnel Interface (e.g "**Interface Tunnel 0**") is configured as either GRE or VTI mode. Then, in order to have security protection of the VPN, an IPSEC profile is attached to the Tunnel interface. All traffic that passes through this Tunnel Interface is encrypted and placed in the VPN. Static or Dynamic routing is used to move traffic towards this Tunnel Interface in order to pass through the VPN tunnel. As we've said above, Route-Based VPNs are based either on GRE or VTI technologies. Let's start with GRE based VPNs.

1.3.1 VPN using GRE

Generic Routing Encapsulation (GRE) was originally developed by Cisco but later on was standardized and is now being used by many other vendors. GRE encapsulates packets into an extra IP header (with extra IP address and 4-bytes extra GRE header) and sends this new packet across the network. If you have two separated LAN networks with private IP addresses, you can create a GRE VPN tunnel between them over the Internet and allow the two private LAN subnets to communicate. The private IP packets will be encapsulated inside a new GRE IP packet (which will use the public IP address as a new header of the private IP packets) and thus the two private LAN subnets can communicate over the Internet.

The diagram below shows a simple Site to Site VPN using GRE encapsulation:

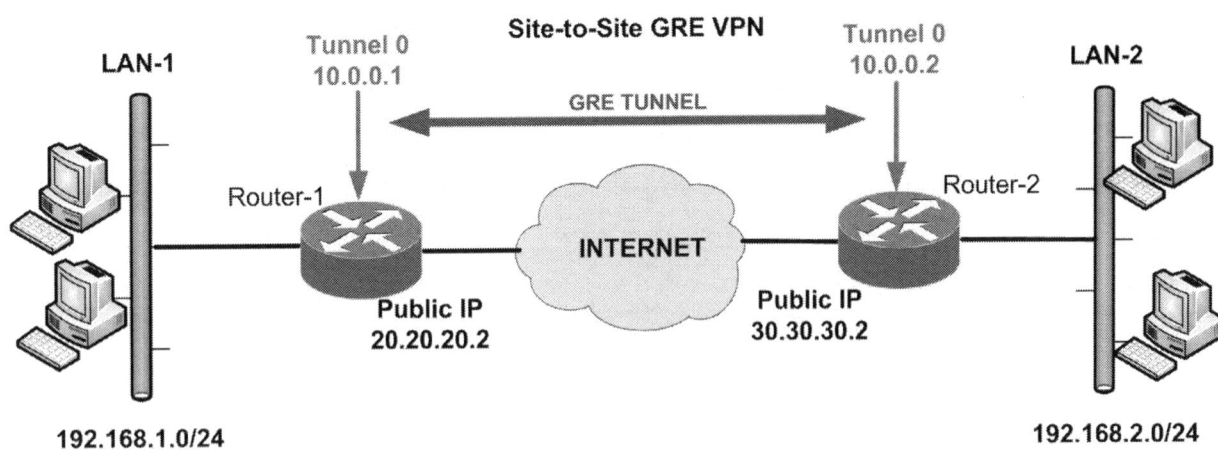

NOTE: GRE is supported only on Cisco Routers. ASA Firewalls do not support GRE VPN.

As shown on the diagram above, the two Routers are connected to the Internet with Public IP addresses (20.20.20.2 and 30.30.30.2). Since the two public IP addresses are reachable via the Internet, you can configure a GRE Tunnel between them, and thus you can allow the two private LAN networks (192.168.1.0 and 192.168.2.0) to communicate between them. You must also configure a Tunnel virtual interface (**Tunnel 0**) on each router which will be used to run the GRE traffic encapsulation. Each Tunnel interface must have a private IP address in the same network range with the other site's Tunnel interface (10.0.0.1 and 10.0.0.2 in example above).

The diagram above shows only two sites. You can configure a Hub-and-Spoke topology also (i.e one Hub Site with two or more Spoke remote sites) but you will need to configure different Tunnel Interfaces (Tunnel 0, Tunnel 1 etc) in order to have a point-to-point GRE tunnel between the Hub and each Spoke. You can find all configuration details for GRE VPNs in Sections 2.2.1, 2.2.2.

1.3.1.1 GRE Vs IPSEC

The above diagram and description looks similar with site-to-site IPSEC VPN functionality. However, one of the main differences between GRE and traditional IPSEC is that GRE VPN does NOT provide encryption or any other security to the packets compared to IPSEC VPN. The best option for GRE VPN is to combine it with IPSEC. This means that we can protect the GRE Tunnel inside an IPSEC Tunnel, thus providing security as well (see diagram below):

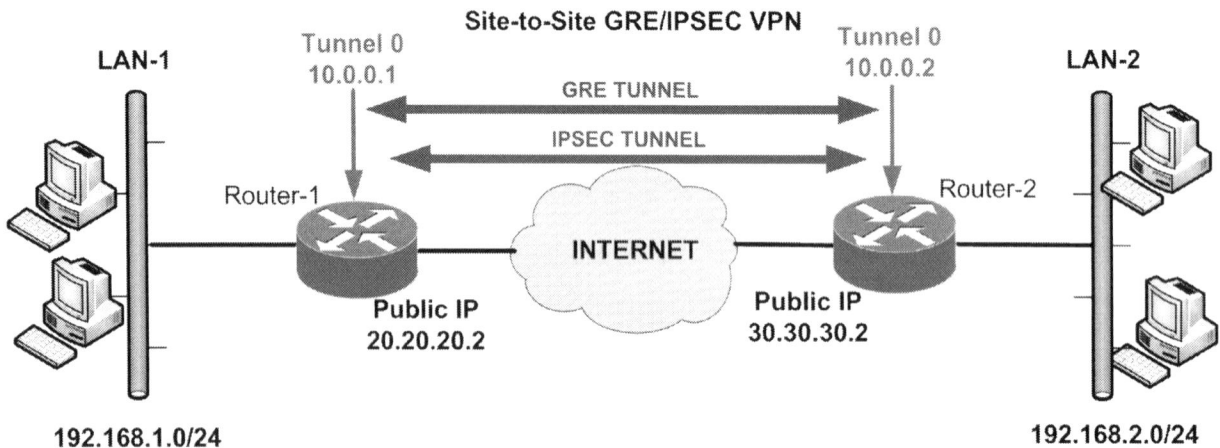

NOTE:

The scenario shown above is an example of "**Route-Based VPN**" which we mentioned in section 1.1 above. We will see more route-based VPNs later in the section of Virtual Tunnel Interfaces (VTI).

Another difference between GRE and traditional IPSEC is that with GRE VPN you can pass multicast and other non-IP traffic inside the tunnel. This is not supported with traditional IPSEC VPN (policy-based IPSEC). Only IP unicast traffic can pass through a traditional IPSEC tunnel. The diagram below shows an implementation of GRE VPN with routing protocol communication between two sites:

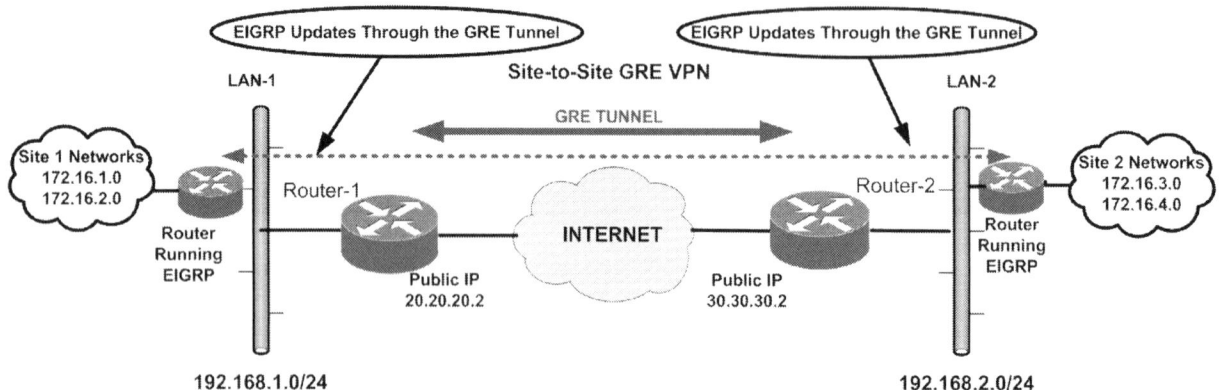

As shown above, Site1 and Site2 have several internal networks. With GRE tunnel in place you can run routing protocols (such as EIGRP or OPSF) between the two sites in order to advertise all internal networks from one site to the other. EIGRP or OSPF use multicast for routing updates communication. Multicast can pass with no problems through the GRE tunnel. Moreover, you can also apply IPSEC protection on top of GRE for protecting everything, and thus you can have the best of both worlds (GRE and IPSEC combined).

The Table below illustrates a comparison between traditional IPSEC and GRE VPNs.

	Traditional IPSEC VPN (policy-based VPN)	**GRE VPN (route-based VPN)**	**Combination IPSEC/GRE (route-based IPSEC VPN)**
Security	Strong	None	Strong
Data Protocols Supported	Only IP Unicast Traffic	Multicast and several non-IP protocols supported	Multicast and several non-IP protocols supported
Cisco Devices Support	ASA Firewalls, Cisco Routers	Only on Cisco Routers	Only on Cisco Routers

NOTE:

The **non-IP protocols** supported by GRE include IPX, SNA, Appletalk, DECNet, Banyan Vines etc.

1.3.2 VPN using Virtual Tunnel Interface (VTI)

The second type of Route-Based VPN that we will talk about is Virtual Tunnel Interface. VTI is a special Layer3 Interface type (supported only on Cisco Routers) and is used to create Route-Based VPNs. It is very similar with GRE with some differences as we will see later. VTI is always configured with IPSEC protection. All traffic that passes through the VTI interface is encrypted with IPSEC (similar with GRE-combined-with-IPSEC example before).

There are two types of VTI:

- **Static VTI (SVTI)**: Very similar with point-to-point GRE VPN implementation using tunnel interfaces. Used mainly for few sites to create site-to-site VPNs.
- **Dynamic VTI (DVTI)**: It uses **Virtual Templates** similar with legacy dial-in implementations. Very useful in Hub-and-Spoke deployments where you have several spoke sites. The Hub router can use a single DVTI and the remote Spoke sites can use a Static VTI to connect to the Hub. New spokes can be added without changing the HUB configuration.

1.3.2.1   Static VTI

The diagram below shows a simple Static Virtual Tunnel Interface (SVTI) implementation.

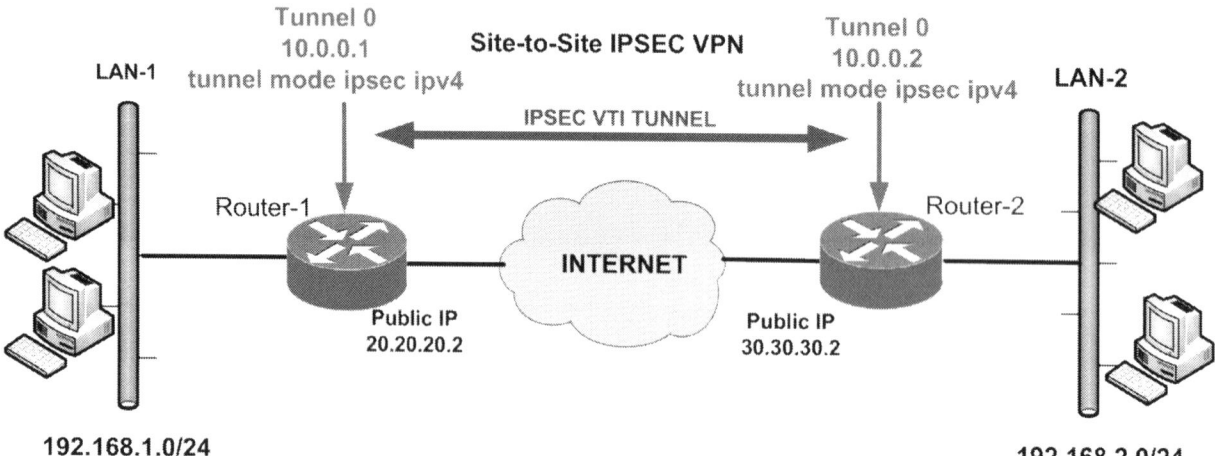

As you can see from the diagram above, it's very similar with GRE VPN using Tunnel Interfaces. The command "**tunnel mode ipsec ipv4**" configures the Tunnel Interface as VTI which can support native IPSEC. The default mode of a Tunnel interface is GRE. By configuring the Tunnel interface as VTI, we eliminate the extra 4-bytes overhead encapsulation used by GRE. However, the VTI interface supports multicast and IP unicast traffic only, compared with GRE which supports also several non-IP protocols in addition to multicast.

The diagram above shows only two sites. You can configure a Hub-and-Spoke topology also (i.e one Hub Site with two or more spoke remote sites) but you will need to configure different Tunnel Interfaces (Tunnel 0, Tunnel 1 etc) in order to have a point-to-point SVTI tunnel between the Hub and each Spoke. You can find all configuration details for Static VTI in Section 2.2.3

The Table below illustrates the main differences between GRE and VTI VPNs.

	GRE VPN	**VTI VPN**
Security	Strong (with IPSEC)	Strong (with IPSEC)
Data Protocols Supported	Multicast, Unicast, and several non-IP protocols supported	Multicast and Unicast Only
Overhead	Extra 4-bytes needed for GRE	No extra overhead
Cisco Devices Support	Only on Cisco Routers	Only on Cisco Routers

1.3.2.2 *Dynamic VTI*

A Dynamic VTI (DVTI) was originally used for creating remote-access VPNs using the EazyVPN feature. However, in newer router IOS versions, DVTI is suitable for creating scalable and easy to manage Hub-and-Spoke topologies as shown below.

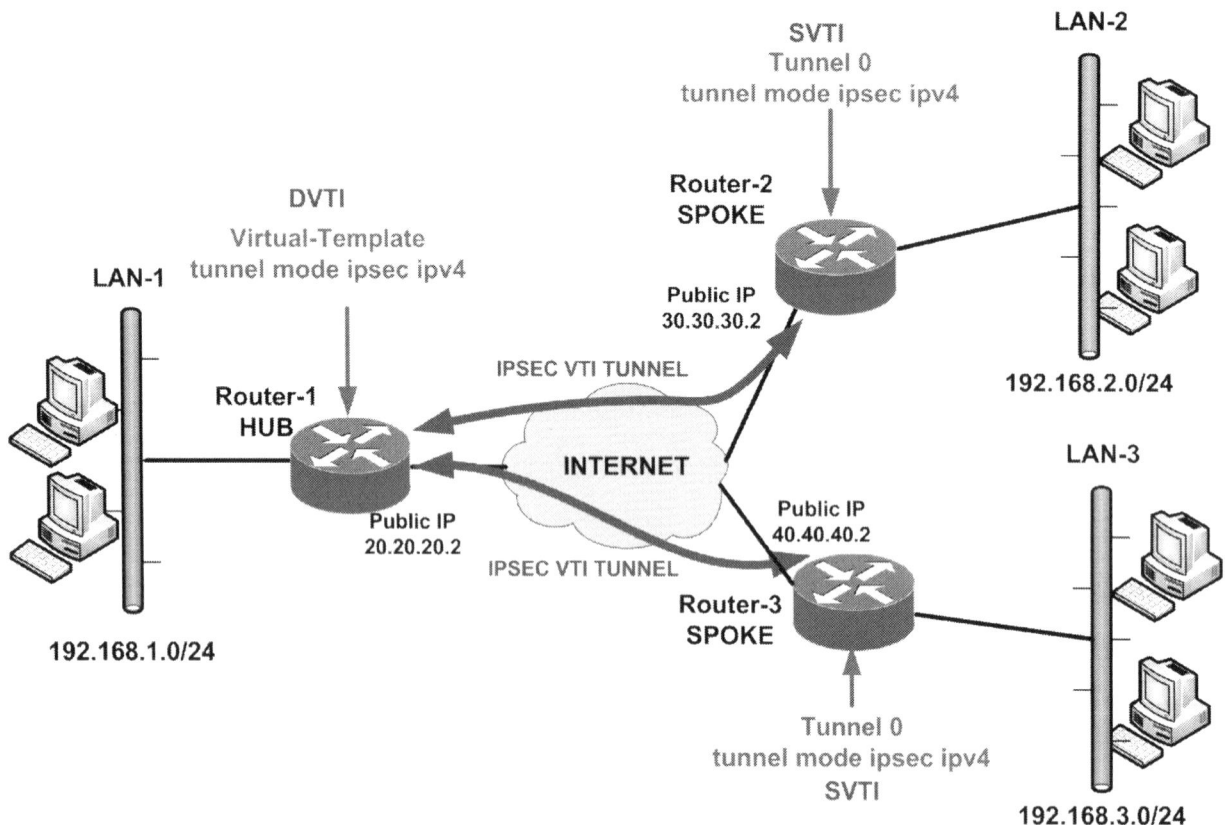

A DVTI requires minimal configuration on the HUB router. A single "**Virtual-Template**" interface is configured on the Hub and an IPSEC security profile can be attached on this interface for protection. The remote Spoke branch sites can use Static VTI interfaces (Tunnel Interface) and create dynamic IPSEC VTI tunnels with the HUB. Each Spoke-to-Hub tunnel creates a dynamic "virtual-access" tunnel interface on the HUB which is cloned from the Virtual-Tunnel interface. If this sounds confusing it will get clear when we see the configuration details in Section 2.2.4.

The configuration of the central HUB site does not need to change when new Spoke sites are added in the topology. The "Virtual-Template" concept was originally used in legacy dial-up networks were multiple remote dial-up clients could connect to a central dial-up router.

Dynamic routing protocols (such as OSPF and EIGRP) can be configured on both HUB and Spoke sites thus making the whole topology very scalable and easy to deploy. Through the dynamic routing protocol, all Spoke sites will learn the networks of the other branch Spokes, and therefore the spokes can communicate between each other through the central Hub router (this scenario applies in all Route-Based VPNs). In order to have direct Spoke-to-Spoke communication you need to go with the DMVPN technology which we will describe next.

1.4 Dynamic Multipoint VPN (DMVPN)

DMVPN is the most scalable and most efficient VPN type. It is used almost exclusively with Hub-and-Spoke topologies where you want to have <u>direct</u> Spoke-to-Spoke VPN tunnels in addition to the Spoke-to-Hub tunnels. This means that Spoke sites can communicate between them directly without having to go through the Hub. DMVPN is supported only on Cisco Routers.

NOTE:

With the previous Route-Based VPNs you can still have Spoke-to-Spoke communication but the traffic has to go through the Hub in order to reach the other Spoke site.

Our discussion on DMVPN will be based on the following diagram:

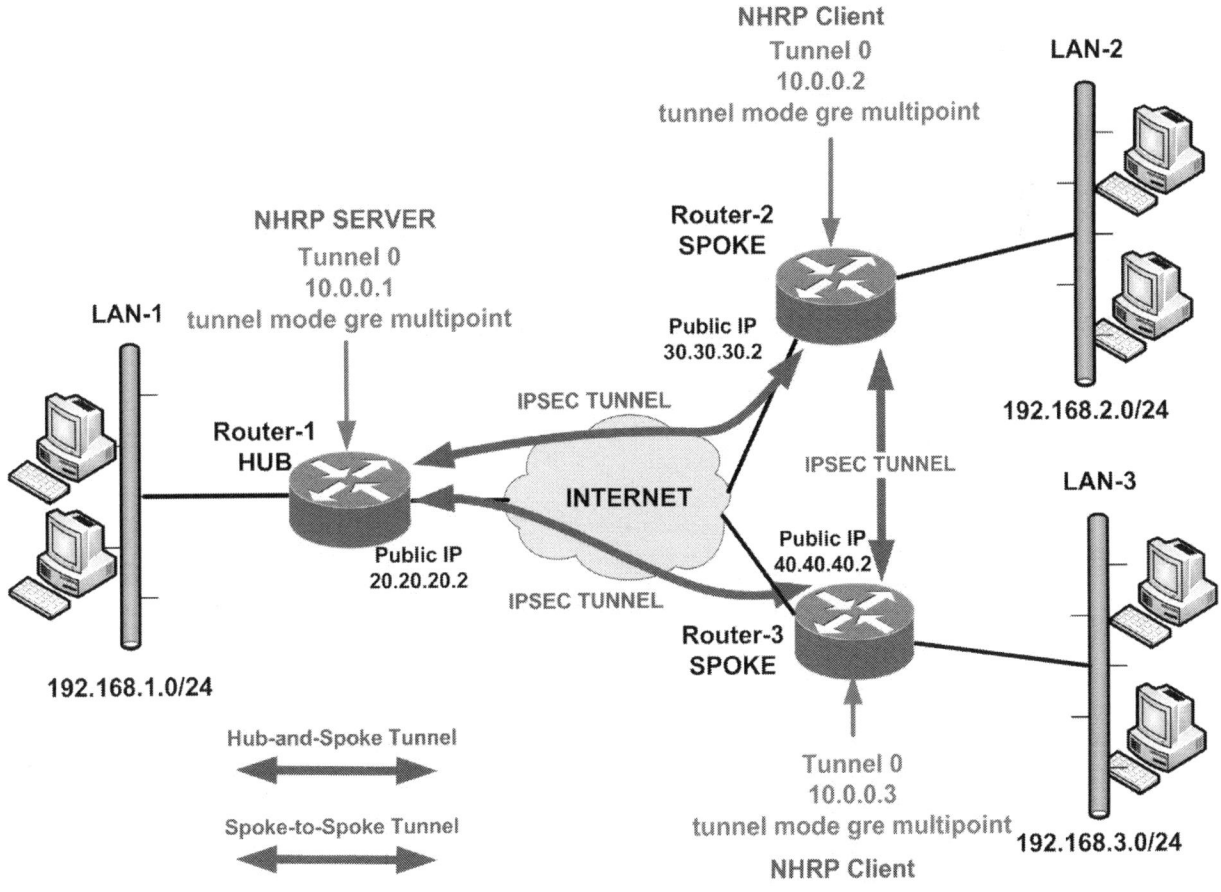

The following are some key points to have in mind about DMVPN:
- Each branch site (Spoke) has a permanent IPSEC Tunnel with the Central site (Hub).
- The Spoke-to-Spoke tunnels are established on demand whenever there is traffic between the Spoke sites. Thereafter, packets are able to bypass the Hub site and use the spoke-to-spoke tunnel directly.
- All tunnels are using Multipoint GRE with IPSEC Protection.
- NHRP (Next Hop Resolution Protocol) is used to map the private IPs of Tunnel Interfaces with their corresponding WAN Public IPs. For example, NHRP will map Tunnel IP 10.0.0.2 (Router-2) with its public WAN IP of 30.30.30.2. Similar mapping happens with Router-3 as well.
- The above NHRP mappings will be kept on the NHRP Server router (HUB). Each Spoke communicates with the NHRP Server (Hub) and registers its public IP address and its private Tunnel Interface IP to the Hub router. Thus, the Hub router will store all mappings for "**Tunnel Interface IP / Public WAN IP**" of all the Spoke sites.
- When a spoke needs to send a packet to a destination (private) subnet on another spoke, it queries the NHRP server in order to learn the public (outside WAN) address of the destination (target) spoke.
- A dynamic routing protocol (e.g EIGRP) is running between all sites, thus advertises all IP addresses (Tunnel private, LAN private) to all other routers.

Example Communication between Router-2 to Router-3
- From our diagram above, Router-2 knows that subnet 192.168.3.0/24 (LAN-3) is reachable via Tunnel IP 10.0.0.3. This is learned via the dynamic routing protocol running between all sites.
- However, Router-2 does not know yet the public IP of Router-3. Thus, it queries the NHRP Server (Hub router) in order to learn the public IP mapping for Tunnel IP 10.0.0.3.
- The NHRP Server will reply that 10.0.0.3 corresponds to public IP 40.40.40.2 (WAN of Router-3). Note that the public WAN IP of Router-3 can be a dynamically assigned IP (no need to be static IP).
- Thus a GRE/IPSEC tunnel will be created dynamically between Router-2 and Router-3.

- Now Router-2 will encapsulate all private IP traffic from its own LAN (192.168.2.0/24 network) into this new GRE/IPSEC Tunnel and send it to Router-3. Therefore, we have now direct tunnel communication between LAN-2 and LAN-3.

The Table below shows a comparison of the main similarities and differences between DMVPN and the other Route-Based VPNs that we've described before (i.e GRE and VTI).

	DMVPN	**Route-Based VPNs (GRE and VTI)**
Security	Strong (when using IPSEC)	Strong (when using IPSEC)
Multicast Support	Yes	Yes
Dynamic Routing Protocols through the tunnel	Supported	Supported
Scalability	Excellent	Very Good (on DVTI) Not Good (on the other types)
Communication between sites	Direct communication between Spoke-to-Hub and Spoke-to-Spoke	Direct communication between Spoke-to-Hub. Spoke-to-Spoke traffic goes through the Hub.
Configuration Complexity	High	Low
Cisco Devices Support	Only on Cisco Routers	Only on Cisco Routers

You can find all configuration details of DMVPN in Section 2.3

1.5 SSL Based VPNs (WebVPN)

In this Section we will describe the newest remote access VPN functionality supported on Cisco Devices, called SSL Based VPNs. This type is also called WebVPN in Cisco Terminology. SSL based VPNs are used to allow remote users to connect to their central network via a normal Web Browser with SSL encryption.

1.5.1 Types of SSL Based VPNs

There are mainly two types of SSL VPNs supported by Cisco devices:

- **Clientless Mode WebVPN:** This is the first implementation of SSL WebVPN supported. It lets users establish a secure remote access VPN tunnel using just a Web browser. There is no need for a software or hardware VPN client. However, only limited applications can be accessed remotely.
- **AnyConnect WebVPN:** A special Java based client is installed on the user's computer providing an SSL secure tunnel to the central site. Provides full network connectivity (similar with IPSec Remote Access client). All applications and network resources at the central site can be accessed remotely.

1.5.2 Comparison between SSL VPN Technologies

In this book we will focus only on AnyConnect WebVPN. I decided not to bother with the Clientless WebVPN because I believe that the benefits of using AnyConnect instead of Clientless are much more. To justify what I'm saying, let's see the differences between the two WebVPN modes and I'm sure you will understand why I focus only on AnyConnect!

- **Clientless WebVPN** does not require any VPN client to be installed on user's computer. It uses a normal web browser. By pointing the browser to **https://[outside address of ASA]** the user authenticates with the central office device and gets access to a Web Portal. Through this Web Portal, the user can then access a limited number of internal applications. Specifically, only internal Web applications (HTTP, HTTPs), email servers (POP3, SMTP, IMAP), Windows file shares and a small number of TCP legacy applications (e.g Telnet) can be accessed. That is, there is no full network connectivity with Clientless WebVPN.

- **AnyConnect WebVPN,** on the other hand, provides FULL network connectivity to the remote user who connects with a web browser and a Java client gets installed on his/her computer. The ASA firewall or Router, working as AnyConnect WebVPN server, assigns an IP address to the remote user and attaches the user to the network. Thus, all IP protocols and applications function across the SSL VPN tunnel without any problems. For example, a remote user, after successfully connected with AnyConnect VPN, can open a Remote Desktop connection and access a Windows Terminal Server inside the central network. Although a special Java-based client is required to be installed on the user's computer, this client can be supplied dynamically to the user from the ASA or Router device. The user can connect with a browser to the ASA or Router and download the Java client on demand. The Java client can remain installed or even get removed from the user's desktop when disconnected from the VPN. This Java client is small in size (around 3MB) and is stored on the ASA or Router flash memory. The newest Anyconnect product is called now "**Cisco Anyconnect Secure Mobility Client**" and from version 3.x and above it supports both SSL and IKEv2 IPSEC VPNs.

1.5.3 Overview of AnyConnect VPN operation:

The diagram below shows a network topology with ASA and a remote user with AnyConnect VPN.

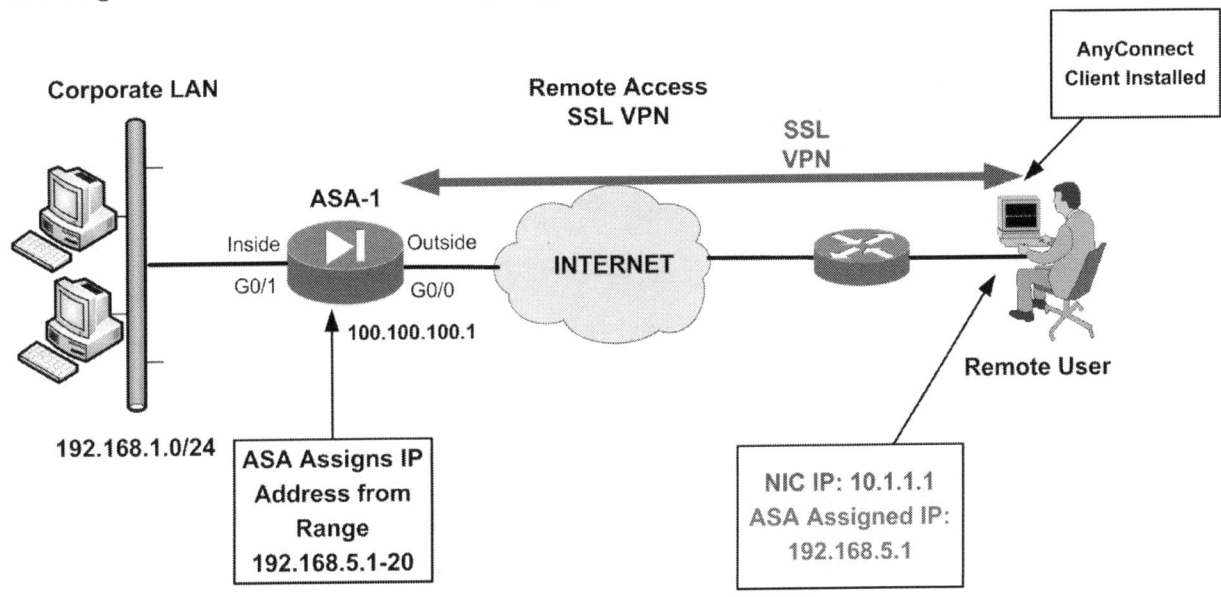

From the diagram above, the ASA firewall is configured as AnyConnect WebVPN server. A remote user has access to the Internet and has an IP address on his/her laptop interface card of 10.1.1.1 (NIC IP). The user can also be behind a router doing NAT/PAT and have his private IP address translated to a public IP by the NAT router. When the remote user connects and successfully authenticates to the ASA with the AnyConnect client, the ASA will assign an internal IP address to the user from a preconfigured IP address range (in our example above, this address range is 192.168.5.1 up to 192.168.5.20). From the diagram above, the ASA assigns IP 192.168.5.1 to the remote user. This means that the remote user is virtually attached to the corporate LAN behind the ASA firewall.

The operation overview described above assumes that the AnyConnect client is already installed on the user's laptop. Let's see below the available options how to initially install the AnyConnect client.

There are two Initial Installation options for AnyConnect client:
- Using clientless WebVPN portal.
- Manual installation by the user

Using the clientless Web portal, the user first connects and authenticates to the ASA with a secure web browser and the Java Anyconnect client is automatically downloaded and installed on the user's computer (the user can also click the "AnyConnect" Tab on the WebVPN portal to download the client). This means that the Java client (**.pkg extension**) is already stored on the ASA flash memory by the administrator (you need to download it from Cisco site). This is the preferred method in my opinion because it automates the distribution of the client to the remote users.

With the manual installation method, the network administrator must download the appropriate Java client (Microsoft MSI package installer or one of the other OS versions) from Cisco site and provide the file to the users for manual installation on their laptop. With this method, the user does not need to log in via clientless mode to start the SSL VPN tunnel. Instead, the users can start up the AnyConnect client manually from their desktop and provide their authentication credentials. You can find all configuration details in Section 3.2.1.

1.6 Practical Applications for each VPN Type

Since the aim of this book is to be a practical reference of Cisco VPN implementations, I would like to discuss a few practical applications that each VPN type described above can be used. I won't go into much detail. Just some bullet points will suffice. Let's start with Policy-Based VPN:

1.6.1 Policy-Based (Traditional IPSEC) VPN Applications

- Because IPSEC is standardized, traditional IPSEC implementations are supported on wide range of devices. Both Cisco Routers and Cisco ASA support traditional IPSEC.
- If you want to create VPNs between devices from different vendors, then policy-based IPSEC is the best option.
- Also, if you have a mixture of Cisco Routers and Cisco ASA firewalls in your topology, then again traditional IPSEC is your only option.
- Keep in mind that only unicast traffic is supported. If you want to have routing protocols or other multicast traffic through the tunnel, then traditional IPSEC will not work.
- For Hub-and-Spoke topologies, traditional IPSEC does not scale well. I would say from experience that if you have more than 8-10 remote sites connected to a central hub then maybe a different VPN type will be more appropriate.
- For Hub-and-Spoke topologies, if you have remote sites which receive a dynamic WAN public IP address (e.g via DHCP or PPPoE) then you must configure a dynamic crypto map on the central Hub device.
- If you are using NAT in order to provide Internet access to your internal LAN, then you must configure special NAT exemption for the traffic that will go inside the IPSEC Tunnel.
- If you have Cisco ASA devices in your network, then traditional IPSEC is your only option.

1.6.2 Route-Based GRE VPN Applications

- If you want to have a VPN network that will support IP unicast, multicast (e.g routing protocols), and several non-IP protocols, then GRE protocol is your only option.
- Keep in mind that GRE does not provide encryption, so you must use it in conjunction with IPSEC if you need secure communications.
- If you have Cisco ASA firewalls in your network, then GRE VPN is not supported. Only Cisco Routers support GRE.
- For Hub-and-Spoke topologies, point-to-point GRE tunnels are not very scalable since you need to configure a different "Tunnel Interface" for each point-to-point tunnel link (i.e for each Hub-to-Spoke tunnel). In my opinion, if you have more than 10 remote branches connected to a central site, then choose a more scalable solution such as Dynamic VTI or DMVPN. However, these two technologies (DVTI and DMVPN) support only unicast and multicast traffic in contrast with GRE which supports also non-IP protocols additionally.
- In Hub-and-Spoke topologies, you can have communication between remote branches (i.e spoke-to-spoke traffic) via the Hub central router (no direct spoke-to-spoke though). You will need to run a dynamic routing protocol (e.g EIGRP, OSPF) between all sites in order to distribute routes of the private LAN networks to all sites.
- If you want to support non-IP traffic (such as IPX, Appletalk etc) then GRE is your only option.
- If you have sites that receive dynamic outside IP address from the ISP, then point-to-point GRE tunnels will not work since you have to specify a source and destination IP address on the Tunnel Interfaces. You need to configure multipoint GRE (with DMVPN) or DVTI in order to support branches that have dynamic outside public IP.

1.6.3 Route-Based VTI VPN Applications

- If you don't need to support non-IP protocols, but you want to have multicast in your VPN network, then Virtual Tunnel Interface (VTI) VPNs offer a preferred option compared to GRE.
- VTI tunnels use less overhead than GRE, so you have more efficient bandwidth utilization.
- If you have Cisco ASA firewalls in your VPN topology, then VTI is not supported. Only Cisco Routers support VTI.
- For Hub-and-Spoke topologies with not a lot of remote branches (e.g up to 10), you can use Static VTI VPNs.
- For Hub-and-Spoke topologies with many remote branches (e.g more than 10), a more scalable option is to use Dynamic VTI on the Hub site and Static VTI on the remote sites.
- In Hub-and-Spoke topologies, you can have communication between remote branches (i.e spoke-to-spoke traffic) via the Hub central router (no direct spoke-to-spoke though). You will need to run a dynamic routing protocol (e.g EIGRP, OSPF) between all sites in order to distribute routes of the private LAN networks to all sites.
- If you have sites that receive dynamic outside IP address from the ISP, then you need to use DVTI on the Hub Router and SVTI on the remote branch routers.

1.6.4 Dynamic Multipoint VPN Applications

- If you have a large VPN network topology with a central Hub Site and numerous (tens or hundreds) remote branch sites that need to communicate directly between them, then DMVPN is the best option.
- Remote branch sites have a permanent IPSEC VPN tunnel with the Hub site, and also direct IPSEC VPN tunnels are created between branch sites on demand (i.e whenever there is traffic between branches).
- Supports unicast and multicast traffic in the VPN tunnels.
- Supported only on Cisco Routers.

- DMVPN is also ideal if you have VoIP running in your sites. The direct branch-to-branch communication allowed by DMVPN reduces latency and jitter thus improving network performance and VoIP quality.
- DMVPN supports also remote branch sites that receive a dynamic public IP address from the ISP. No need to have static public IP address for the remote branches.

Chapter 2   VPN Configuration on Cisco Routers

In this Chapter we are getting into more technical details of VPN configuration. We will see how the various VPN categories we've described in Chapter 1 are actually implemented on Cisco devices. This Chapter will focus on VPN configuration on Cisco Routers and the next Chapter will be about VPN configuration on Cisco ASA firewalls.

2.1   Policy-Based VPN Configuration on Cisco Routers

2.1.1 Site-to-Site IPSEC VPN

This is the simplest form of traditional IPSEC VPN configuration. Since this is the first configuration we are going to describe, and because it's the base of other IPSEC implementations that we are going to see later on, we will describe this configuration in step-by-step details.

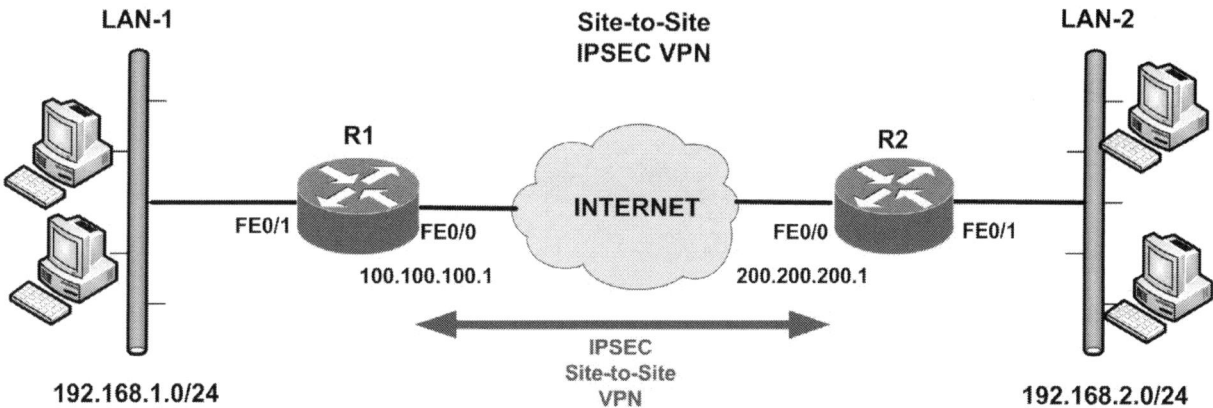

In our example network topology shown above, we have two sites that we want to connect through the Internet with an IPSEC VPN tunnel. LAN-1 and LAN-2 will be able to communicate securely over the Internet. Both sites have a static public IP address assigned by the ISP. In another scenario later we will see also a case in which one of the sites has dynamic public IP from the ISP.

As we have described in section 1.2.2 above in "**How IPSEc Works**", there are five steps in the operation of IPSEc. Next we will describe the configuration commands needed for each step in order to set up the VPN. All configuration commands below refer to the network diagram shown above.

- **STEP 1: Configure Interesting Traffic**

We need first to define the **Interesting Traffic**, that is, traffic that we want to encrypt. Using Access-Lists (**Crypto ACL**) we can identify which traffic flow must be encrypted. In our example diagram above, we want all traffic flow between private networks 192.168.1.0/24 and 192.168.2.0/24 to be encrypted. However, all other traffic from LAN-1 or LAN-2 to the Internet will not pass through the VPN tunnel.

Configuring "Interesting Traffic" for encryption is one of the reasons why this type of VPN is also called "Policy-Based VPN": A selected subset of traffic passing through an interface (usually the WAN interface) is encrypted and inserted in the VPN tunnel according to a predefined policy which is implemented with the "Crypto ACL" and several other parameters.

Router R1:

R1(config)# ip access-list extended VPN-ACL
R1(config-ext-nacl)# permit ip 192.168.1.0 0.0.0.255 192.168.2.0 0.0.0.255

Router R2:

R2(config)# ip access-list extended VPN-ACL
R2(config-ext-nacl)# permit ip 192.168.2.0 0.0.0.255 192.168.1.0 0.0.0.255

Notice that we have to configure the exact mirror access-list for each Router participating in the IPSEc VPN. The Crypto ACL needs to identify only outbound traffic on each router. The **permit** statement in the ACL means that the specific traffic flow will be encrypted and transported through the VPN tunnel.

NAT Exemption

One important issue to consider is the case of using NAT on the router for Internet access. Because IPSEc does not work with NAT, we need to <u>exclude</u> the traffic to be encrypted from the NAT operation. This means in our example that the Interesting Traffic in the Crypto ACL must not be translated (must be excluded from NAT).

The configuration below shows how to exclude VPN tunnel traffic from NAT operation. If you don't use NAT, then the configuration below is not needed.

Router R1:

R1(config)# ip access-list extended NAT-ACL

R1(config-ext-nacl)# deny ip 192.168.1.0 0.0.0.255 192.168.2.0 0.0.0.255 ← Exclude traffic from LAN1 to LAN2 from NAT operation

R1(config-ext-nacl)# permit ip 192.168.1.0 0.0.0.255 any ← Allow all other traffic from LAN-1 to be NATed

! Enable the NAT functionality on FE0/1 (inside) and FE0/0 (outside) interfaces

R1(config)# ip nat inside source list NAT-ACL interface FastEthernet0/0 overload

R1(config)# interface FastEthernet0/0

R1(config-if)# ip nat outside

R1(config)# interface FastEthernet0/1

R1(config-if)# ip nat inside

Router R2:

R2(config)# ip access-list extended NAT-ACL

R2(config)# deny ip 192.168.2.0 0.0.0.255 192.168.1.0 0.0.0.255 ← Exclude traffic from LAN2 to LAN1 from NAT operation

R2(config)# permit ip 192.168.2.0 0.0.0.255 any ← Allow all other traffic from LAN-2 to be NATed

! Enable the NAT functionality on FE0/1 (inside) and FE0/0 (outside) interfaces

R2(config)# ip nat inside source list NAT-ACL interface FastEthernet0/0 overload

R2(config)# interface FastEthernet0/0

R2(config-if)# ip nat outside

R2(config)# interface FastEthernet0/1

R2(config-if)# ip nat inside

- STEP 2: Configure Phase 1 (ISAKMP)

Phase 1 of the IPSEc operation is used to establish a secure communication channel for further data transmission. In Phase 1, VPN peers exchange shared secret keys, authenticate each other, negotiate IKE security policies etc. In this Phase we configure an **isakmp policy** which MUST match the policy configured on the other peer(s). This **isakmp policy** tells the other peer(s) what security parameters must be used in the VPN (e.g encryption protocol, hash algorithm, authentication method, Diffie Hellman Group (DH), lifetime threshold for the tunnel etc).

Several isakmp policies can be configured to match different requirements from different IPSEc peers. The priority number uniquely identifies each policy. The lower the priority number, the higher the priority will be given to the specific policy.

The following example parameters can be used to create a good isakmp policy:
- Encryption **3des** or **aes**
- Hash **md5** or **sha**
- Authentication **pre-share**
- Diffie-Helman Group **2 or 5**

Router R1:

R1(config)# crypto isakmp policy 1 ← policy number 1 is created. You can have multiple policies.

R1(config-isakmp)# encryption 3des ← use 3DES for encr. AES or DES options available also.

R1(config-isakmp)# hash md5 ← use MD5 for hash. SHA option available also.

R1(config-isakmp)# authentication pre-share ← use pre-shared key for authentication

R1(config-isakmp)# group 2 ← use Diffie-Helman Group 2

R1(config-isakmp)# exit

R1(config)# crypto isakmp key secretkey address 200.200.200.1 ← define the pre-shared key ("secretkey") for authentication with the remote peer with IP 200.200.200.1

Router R2:

R2(config)# crypto isakmp policy 1 ← policy number 1 is created. You can have multiple policies.

R2(config-isakmp)# encryption 3des ← use 3DES for encr. AES or DES options available also.

R2(config-isakmp)# hash md5 ← use MD5 for hash. SHA option available also.

R2(config-isakmp)# authentication pre-share ← use pre-shared key for authentication

R2(config-isakmp)# group 2 ← use Diffie-Helman Group 2

R2(config-isakmp)# exit

R2(config)# crypto isakmp key secretkey address 100.100.100.1 ← define the pre-shared key ("secretkey") for authentication with the remote peer with IP 100.100.100.1

- STEP 3: Configure Phase 2 (IPSEc)

After a secured tunnel is established in Phase 1, the next step in setting up the VPN is to negotiate the IPSEc security parameters that will be used to protect the data and messages within the tunnel. This is achieved in Phase 2 of the IPSEc. In this Phase the following functions are performed:

- Negotiation of IPSEc security parameters and IPSEc **transform sets**.
- Establishment of IPSEc SAs.
- Renegotiation of IPSEc SAs periodically to ensure security.

The ultimate goal of IKE Phase 2 is to establish a secure IPSEc session between peers. Before that can happen, each pair of endpoints negotiates the level of security required (encryption and authentication algorithms for the session). Rather than negotiate each encryption and authentication protocol individually, the protocols are grouped into sets, called **transform sets**. IPSEc transform sets are exchanged between peers and they must match between peers in order for the session to be established.

The following encryption and authentication protocols are supported in most IOS with crypto functionality and can be used in transform sets:

Transform	Description
esp-des	ESP transform using DES cipher (56 bits)
esp-3des	ESP transform using 3DES cipher (168 bits)
esp-aes	ESP transform using AES-128 cipher
esp-aes 192	ESP transform using AES-192 cipher
esp-aes 256	ESP transform using AES-256 cipher
esp-md5-hmac	ESP transform using HMAC-MD5 authentication
esp-sha-hmac	ESP transform using HMAC-SHA authentication
esp-none	ESP with no authentication
esp-null	ESP with null encryption

The following guidelines can be useful when choosing transform protocols:

- For providing data confidentiality (encryption), use an ESP encryption transform such as the first 5 in the list above.
- Highly recommended to use also an ESP authentication transform by choosing MD5-HMAC or SHA-HMAC algorithms (items 6 and 7 on table above).
- SHA is stronger than MD5 but it is slower.

Consider the following example combinations of transform sets:
- **ESP-3DES** and **ESP-MD5-HMAC** for strong encryption and authentication.
- **ESP-AES** and **ESP-SHA-HMAC** for stronger encryption and authentication.

After configuring a transform set on both Router peers, we need to configure a **crypto map** which combines all Phase 2 IPSEc parameters. This crypto map is then attached to the VPN termination interface (usually the WAN of the router) on which the IPSEc will be established.

Let's see now the Phase 2 configuration on the two routers:

Router R1:
R1(config)# crypto ipsec transform-set TRSET esp-3des esp-md5-hmac ← Configure a transform set with 3DES encryption and MD5-HMAC for authentication.
R1(cfg-crypto-trans)# exit

R1(config)# crypto map VPNMAP 10 ipsec-isakmp ← The crypto map will combine all Phase2 parameters
R1(config-crypto-map)# set peer 200.200.200.1 ← The other site of the VPN
R1(config-crypto-map)# set transform-set TRSET ← Transform TRSET configured before.
R1(config-crypto-map)# match address VPN-ACL ← Crypto ACL from Step1.
R1(config-crypto-map)# exit

!Attach the crypto map above to the WAN outside interface (FE0/0) of the router
R1(config)# interface FastEthernet0/0
R1(config-if)# crypto map VPNMAP

Router R2:

R2(config)# **crypto ipsec transform-set TRSET esp-3des esp-md5-hmac** ← Configure a transform set with 3DES encryption and MD5-HMAC for authentication.

R2(cfg-crypto-trans)# **exit**

R2(config)# **crypto map VPNMAP 10 ipsec-isakmp** ← The crypto map will combine all Phase2 parameters

R2(config-crypto-map)# **set peer 100.100.100.1** ← The other site of the VPN

R2(config-crypto-map)# **set transform-set TRSET** ← Transform TRSET configured before.

R2(config-crypto-map)# **match address VPN-ACL** ← Crypto ACL from Step1.

R2(config-crypto-map)# **exit**

!Attach the crypto map above to the WAN outside interface (FE0/0) of the router

R2(config)# **interface FastEthernet0/0**

R2(config-if)# **crypto map VPNMAP**

NOTE:

The number (10) in the crypto map configuration above indicates a sequence number. You can have multiple entries (sequence numbers) in the same crypto map in cases where you have several IPSEC VPN tunnels terminated on the same router (e.g in Hub-and-Spoke topologies where you have several remote branches connected to the same central Hub router).

- **STEP 4: Verify Encrypted Data Transfer**

With the three steps above we concluded the configuration of a site-to-site IPSEc VPN. An essential step though is to verify that everything is working fine and that our data is actually getting encrypted by the routers. There are two important commands that will help you verify if the tunnel is established and if data is bi-directionally encrypted between the IPSEc peers.

Verify that tunnel is established

The **show crypto isakmp sa** command verifies that the Security Association (SA) is established which means that the tunnel is up and running. Let's see an example output of this command below:

R1# show crypto isakmp sa

dst	src	state	conn-id	slot	status
200.200.200.1	100.100.10.1	**QM_IDLE**	1	0	**ACTIVE**

The important points to observe here are the **State : QM_IDLE** and the **Status: ACTIVE**. These two parameters verify that the IPSEc tunnel is established successfully.

Verify that data is bi-directionally encrypted

The **show crypto ipsec sa** command verifies that data is being encrypted and decrypted successfully by the routers, as shown below:

R1# show crypto ipsec sa

```
interface: FastEthernet0/0
  Crypto map tag: VPNMAP, local addr 100.100.100.1

 protected vrf: (none)
 local ident (addr/mask/prot/port): (192.168.1.0/255.255.255.0/0/0)
 remote ident (addr/mask/prot/port): (192.168.2.0/255.255.255.0/0/0)
 current_peer 200.200.200.1 port 500
  PERMIT, flags={origin_is_acl,}
 #pkts encaps: 11, #pkts encrypt: 11, #pkts digest: 11
 #pkts decaps: 11, #pkts decrypt: 11, #pkts verify: 11
 #pkts compressed: 0, #pkts decompressed: 0
 #pkts not compressed: 0, #pkts compr. failed: 0
 #pkts not decompressed: 0, #pkts decompress failed: 0
 #send errors 1, #recv errors 0
----- {Output Omitted}
```

As you can see above, the output fields "**pkts encrypt**" and "**pkts decrypt**" show indeed that we have packets being encrypted and decrypted bi-directionally.

You can find a complete configuration example of the scenario above in Chapter 4, Section 4.1.1.

2.1.1.1 *Site-to-Site IPSEC VPN with Dynamic IP*

One important scenario to discuss is when we have a branch site which obtains a dynamic public IP from the Internet ISP. This for example can be an ADSL connection using PPPoE.

In our example case below, one of the sites (R1) has a static public IP and the second one (R2) has a dynamic IP. All configurations described before will be the same for the dynamic site which connects to the static site. However, the static site which connects to the dynamic site needs to be configured differently as we will see below.

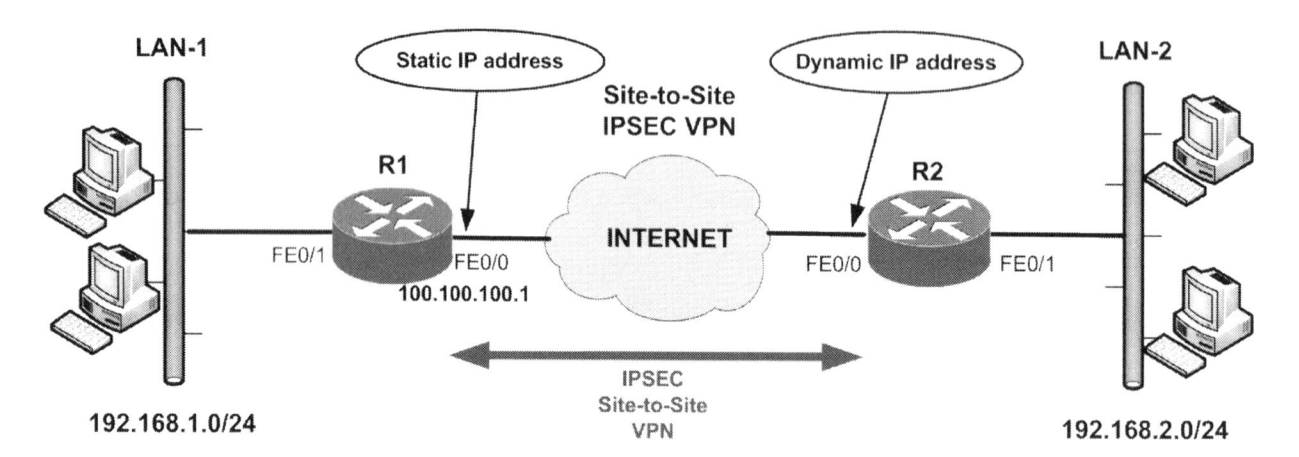

NOTE:

In the above scenario you should keep in mind that the IPSEC VPN tunnel will be established ONLY when LAN-2 initiates traffic to LAN-1. Once the VPN tunnel is established, then LAN-1 can also send traffic to LAN-2.

Let's see the IPSEC configuration for R1 only since IPSEC for R2 will be configured the same as with the previous site-to-site scenario.

Router R1:

The differences are in Phase1 and Phase2 configuration of IPSEC.

Phase1 Configuration

R1(config)# crypto isakmp key secretkey address 0.0.0.0 0.0.0.0

Basically, on Router R1 we configure a wildcard mask (0.0.0.0 0.0.0.0) for the pre-shared key because we don't know the public IP address of the peer site (R2) since it has a dynamic IP. This means that R1 will accept **isakmp** requests from any IP which has the correct pre-shared key authentication password ("**secretkey**" in our example above). It is critical to use a very strong key here.

Phase2 Configuration

As you remember from the previous static Site-to-Site VPN configuration, in Step3 (Phase2 configuration) we had a Crypto map (static crypto map) which was referencing a peer IP address of the remote site. However, since the peer IP address of R2 is not known, we need to configure a "**Dynamic Crypto Map**" which will be used in the "**Static Crypto Map**".

!First create a dynamic crypto map (DYNMAP) which has the Transform Set and Crypto ACL

R1(config)# crypto dynamic-map DYNMAP 10

R1(config-crypto-map)# set transform-set TRSET

R1(config-crypto-map)# match address VPN-ACL

R1(config-crypto-map)# exit

!Then create a static crypto map (VPNMAP) which uses the dynamic map configured before

R1(config)# crypto map VPNMAP 10 ipsec-isakmp dynamic DYNMAP

!Attach the static crypto map (VPNMAP) to the WAN outside interface (FE0/0) of the router

R2(config)# interface FastEthernet0/0

R2(config-if)# crypto map VPNMAP

You can find a complete configuration example of the scenario above in Chapter 4, Section 4.1.2.

2.1.2 Hub-and-Spoke IPSEC VPN

Another popular topology that you will encounter in real world is Hub-and-Spoke. Many Enterprises have a central headquarters site (HUB) and several smaller remote branch sites (SPOKES) which require connectivity to the network resources located in the central site. A VPN with hub-and-spoke configuration is ideal for such topologies.

The diagram below depicts such a topology. We have a central Hub site and two remote Spoke sites, all with static public IP addresses. In Chapter 4 with the complete configuration examples you will find also a scenario with Spoke sites having both a static and a dynamic IP.

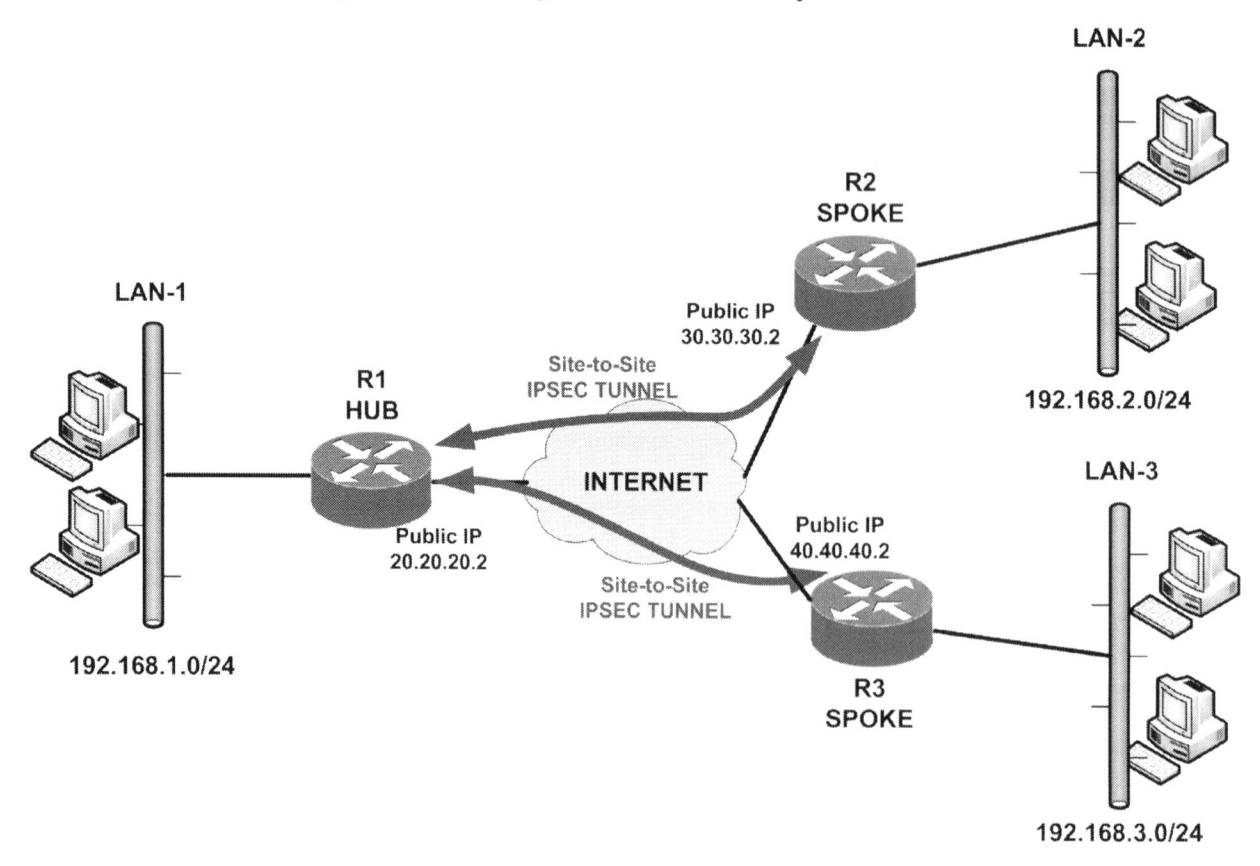

Basically a Hub-and-Spoke VPN network consists of several site-to-site IPSEC VPN tunnels between the Hub and each Spoke site. The configuration for the Spoke remote sites is the same as we've described in "**Site-to-Site IPSEC VPN**" section above so we won't talk about it again. However, the configuration of the Hub site router has a few differences as we will see below:

Router R1 (HUB):

!First identify the Interesting traffic to be encrypted. We need to have two crypto ACLs, one for each Spoke site.

R1(config)# ip access-list extended VPN-TO-REMOTE1
R1(config-ext-nacl)# permit ip 192.168.1.0 0.0.0.255 192.168.2.0 0.0.0.255
R1(config)# ip access-list extended VPN-TO-REMOTE2
R1(config-ext-nacl)# permit ip 192.168.1.0 0.0.0.255 192.168.3.0 0.0.0.255

!Then exclude the VPN Interesting traffic from the NAT operation

R1(config)# ip access-list extended NAT-ACL
R1(config-ext-nacl)# deny ip 192.168.1.0 0.0.0.255 192.168.2.0 0.0.0.255
R1(config-ext-nacl)# deny ip 192.168.1.0 0.0.0.255 192.168.3.0 0.0.0.255
R1(config-ext-nacl)# permit ip 192.168.1.0 0.0.0.255 any

!Now configure the Phase1 isakmp parameters

R1(config)# crypto isakmp policy 1
R1(config-isakmp)# encryption 3des
R1(config-isakmp)# hash md5
R1(config-isakmp)# authentication pre-share
R1(config-isakmp)# group 2
R1(config-isakmp)# exit

!Configure a different pre-shared key for each Spoke site.

R1(config)# crypto isakmp key secretkey1 address 30.30.30.2
R1(config)# crypto isakmp key secretkey2 address 40.40.40.2

!Now Configure Phase2 Transform Set and Crypto Map

R1(config)# crypto ipsec transform-set TRSET esp-3des esp-md5-hmac
R1(cfg-crypto-trans)# exit

!Note below that we have two crypto map entries (VPNMAP 10 & VPNMAP 20) one for each Spoke

R1(config)# crypto map VPNMAP 10 ipsec-isakmp
R1(config-crypto-map)# set peer 30.30.30.2
R1(config-crypto-map)# set transform-set TRSET
R1(config-crypto-map)# match address VPN-TO-REMOTE1
R1(config-crypto-map)# exit

R1(config)# crypto map VPNMAP 20 ipsec-isakmp
R1(config-crypto-map)# set peer 40.40.40.2
R1(config-crypto-map)# set transform-set TRSET
R1(config-crypto-map)# match address VPN-TO-REMOTE2
R1(config-crypto-map)# exit

!Attach the crypto map VPNMAP to the WAN outside interface (FE0/0) of the Hub router

R1(config)# interface FastEthernet0/0
R1(config-if)# crypto map VPNMAP

That's it for the Hub site. You can find a complete configuration example of the scenario above in Chapter 4, Section 4.1.3. Also, Section 4.1.4 shows a scenario with one static and one dynamic spoke.

2.1.3 Remote Access IPSEC VPN

There are two general ways of connecting remote users to a corporate network. One is by using a VPN client software installed on users' computers (IPSEC VPN), and the second one is by using WEB (SSL based) VPN. In this section we will discuss the first type.

In real life, configuring remote access IPSEC VPN on Cisco routers is not very popular (compared to Cisco ASA) but it's a useful scenario to learn. The dedicated Cisco VPN client software uses the IPSEC protocol, so we still need to configure the IPSEC Phase 1 and Phase 2 stages but with some differences compared to the site-to-site IPSEC VPN we've described before. Also, an IP address pool must be configured on the device for dynamically assigning addresses to the remote users. Let's get started with the configuration having as reference the diagram below:

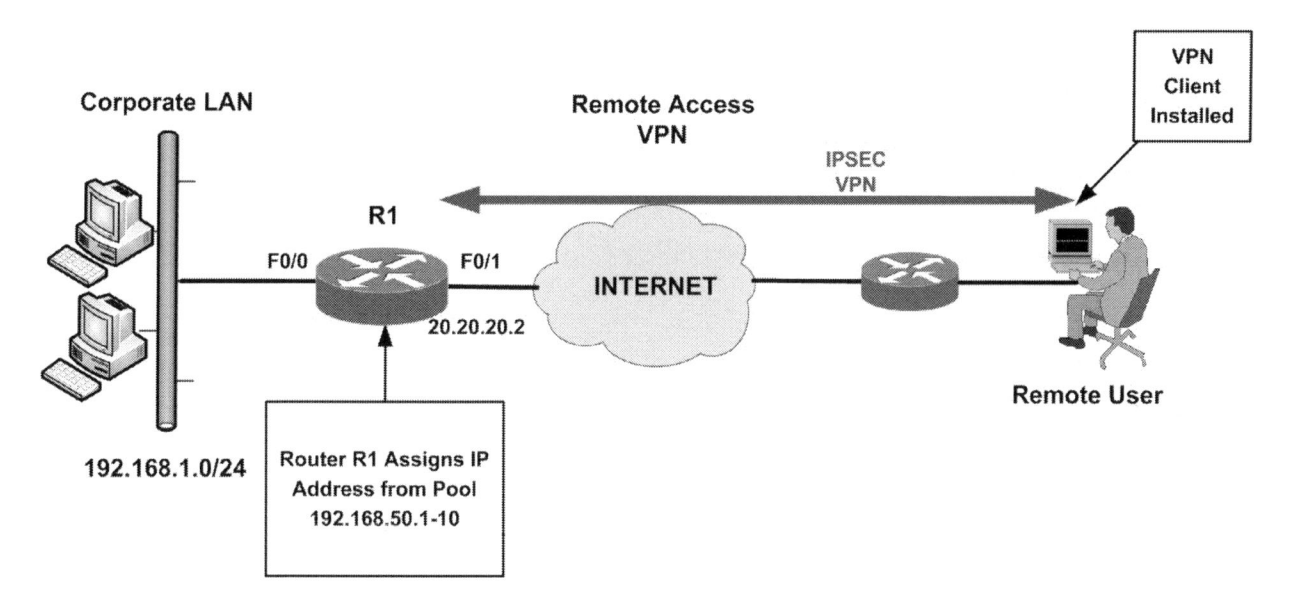

<u>Router R1:</u>

- **STEP 1: Configure VPN Pool**

!First configure an IP address pool that will be used to assign IP addresses to remote users

R1(config)# ip local pool vpnpool 192.168.50.1 192.168.50.10

- **STEP 2: Configure VPN Interesting Traffic and NAT Exemption**

!Identify the Interesting traffic to be encrypted, which is going to be the traffic between LAN (192.168.1.0) and vpnpool (192.168.50.0)

R1(config)# ip access-list extended VPN-ACL
R1(config-ext-nacl)# permit ip 192.168.1.0 0.0.0.255 192.168.50.0 0.0.0.255

!Then exclude the VPN Interesting traffic from the NAT operation

R1(config)# ip access-list extended NAT-ACL
R1(config-ext-nacl)# deny ip 192.168.1.0 0.0.0.255 192.168.50.0 0.0.0.255
R1(config-ext-nacl)# permit ip 192.168.1.0 0.0.0.255 any

! Enable the NAT functionality on FE0/0 (inside) and FE0/1 (outside) interfaces

R1(config)# ip nat inside source list NAT-ACL interface FastEthernet0/1 overload
R1(config)# interface FastEthernet0/1
R1(config-if)# ip nat outside
R1(config)# interface FastEthernet0/0
R1(config-if)# ip nat inside

- **STEP 3: Configure Authentication for Remote Users**

There are a couple of different ways to authenticate and authorize remote users in order to access network resources via the VPN. The "Authentication, Authorization, and Accounting" mechanism (**AAA**) of the router is used for such a task.

The simplest way is to use Local usernames/passwords configured on the Router for authentication and authorization. Another method includes using an external AAA server (Radius or Tacacs) or even Microsoft Active Directory as we will see in the next Chapter. Here we will describe the LOCAL authentication method.

Remote users must be "**authenticated**" first to login to the VPN tunnel, and then must be "**authorized**" to use the network resources. So we must configure the router device for both "login authentication" and "network authorization".

! First configure a LOCAL username/password for each remote user
R1(config)# username vpnuser password strongpassword

! Then enable the AAA mechanism on the router
R1(config)# aaa new-model

! Now configure login authentication and network authorization to use the LOCAL user database
R1(config)# aaa authentication login USERAUTH local

R1(config)# aaa authorization network NETAUTHORIZE local

The above authentication and authorization names (USERAUTH and NETAUTHORIZE) will be used later in IPSEC configuration.

- **STEP 4: Configure IPSEC Phase 1 (isakmp parameters)**

Here, we will see some new features such as "**keyring**", "**isakmp group**", and "**isakmp profile**" which were not used in the site-to-site VPN configuration before.

! First configure an isakmp policy just like the site-to-site IPSEC case.
R1(config)# crypto isakmp policy 10
R1(config-isakmp)# encryption 3des
R1(config-isakmp)# hash md5
R1(config-isakmp)# authentication pre-share
R1(config-isakmp)# group 2

Now we will introduce the concept of "keyring" which is a repository for keeping pre-shared-keys.

!Configure a pre-shared key for the remote access VPN clients.
R1(config)# crypto keyring vpnclientskey
R1(conf-keyring)# pre-shared-key address 0.0.0.0 0.0.0.0 key cisco123

Remote Access users with any IP address (denoted with 0.0.0.0 0.0.0.0) and having a pre-shared key of "cisco123" are allowed to initiate IPSEC Phase 1 connections with the Router. Ofcourse in real world you must use a stronger key.

Now we will configure the policies and parameters that this specific Group of remote users will inherit. Note that we can have multiple Groups of remote users with different policies.

R1(config)# crypto isakmp client configuration group remotevpn ← Group name (see below)
R1(config-isakmp-group)# key cisco123 ← Group Password (see picture below)
R1(config-isakmp-group)# dns 192.168.1.2
R1(config-isakmp-group)# wins 192.168.1.2
R1(config-isakmp-group)# domain mycompany.com
R1(config-isakmp-group)# pool vpnpool ←IP address pool configured before
R1(config-isakmp-group)# acl VPN-ACL ← Split-Tunnel Access List

NOTE1:
The command "**acl VPN-ACL**" sets up split tunneling. That is, traffic will pass through the VPN tunnel only if it matches the access list VPN-ACL. If you remove this ACL, then ALL traffic will have to go through the VPN tunnel (i.e no direct Internet access will be allowed from the location of the remote user).

NOTE2:
The name of the configuration Group that we used above (i.e **remotevpn**) and the key (i.e **cisco123**) must be used in the "Group Authentication" ("Name" /"Password") fields of the VPN Client software (see picture below).

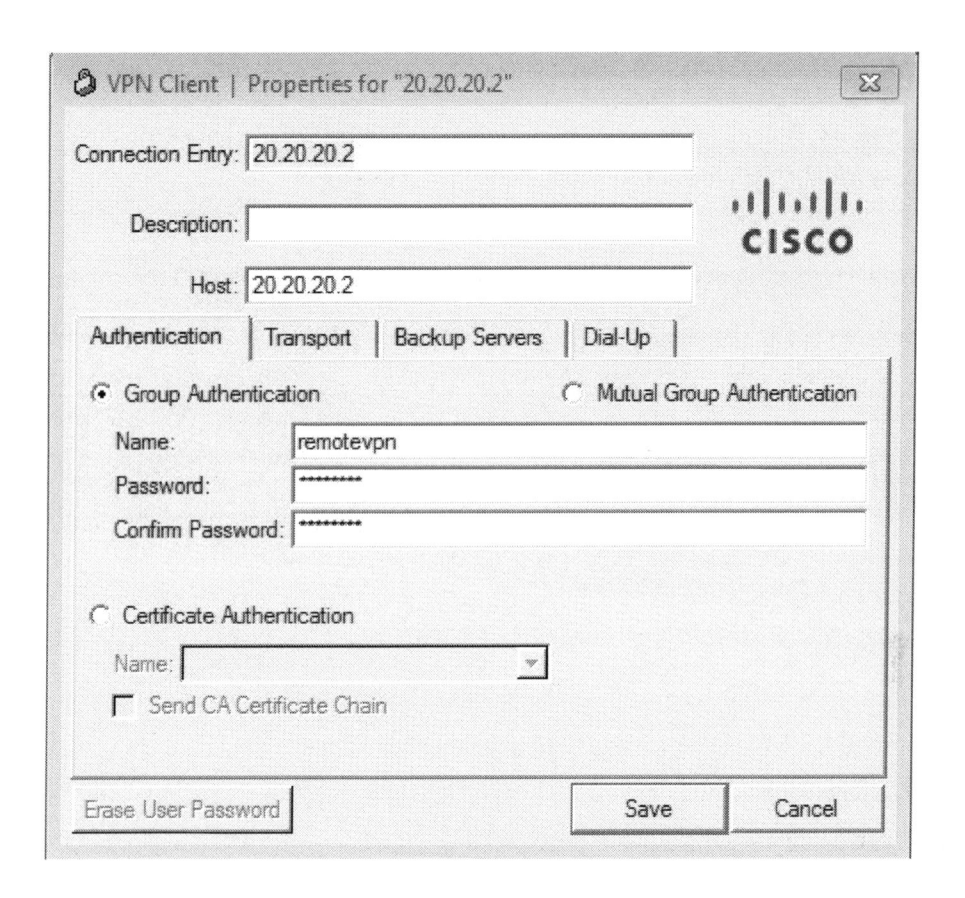

Next we will see another new feature, called "**isakmp profile**". This is a feature introduced to allow modularity and flexibility of IPSEC Phase 1 configurations. This profile will combine all Phase1 elements configured so far and then it will be used later in Phase2 crypto map configuration.

!Configure an isakmp profile.
R1(config)# crypto isakmp profile remoteclients
R1(conf-isa-prof)# description Remote Access VPN clients
R1(conf-isa-prof)# keyring vpnclientskey ← Configured before
R1(conf-isa-prof)# match identity group remotevpn ← Configured before
R1(conf-isa-prof)# client authentication list USERAUTH ← Configured before
R1(conf-isa-prof)# isakmp authorization list NETAUTHORIZE ← Configured before
R1(conf-isa-prof)# client configuration address respond ← Respond to IP address requests from clients.

- **STEP 5: Configure IPSEC Phase 2**

This step has similarities with site-to-site VPN configuration where we had one site with a dynamic IP address. Since the IP address of the remote VPN users will be unknown (dynamic) to the central site router, we have to create a dynamic crypto map.

!Configure Phase2 Transform Set and Crypto Map similar with site-to-site VPN case

R1(config)# crypto ipsec transform-set TRSET esp-3des esp-md5-hmac

R1(cfg-crypto-trans)# exit

!Now create a dynamic crypto map (DYNMAP) which has the Transform Set above and isakmp profile ("remoteclients") configured before.

R1(config)# crypto dynamic-map DYNMAP 10

R1(config-crypto-map)# set transform-set TRSET

R1(config-crypto-map)# set isakmp-profile remoteclients

R1(config-crypto-map)# exit

!Then create a static crypto map (VPNMAP) which uses the dynamic map configured before

R1(config)# crypto map VPNMAP 10 ipsec-isakmp dynamic DYNMAP

!Attach the static crypto map (VPNMAP) to the WAN outside interface (FE0/1) of the router

R2(config)# interface FastEthernet0/1

R2(config-if)# crypto map VPNMAP

That's it for the configuration. After setting up the Cisco VPN client on a computer and connecting with the router, you can click on the Statistics Tab to see the Tunnel Details as shown below:

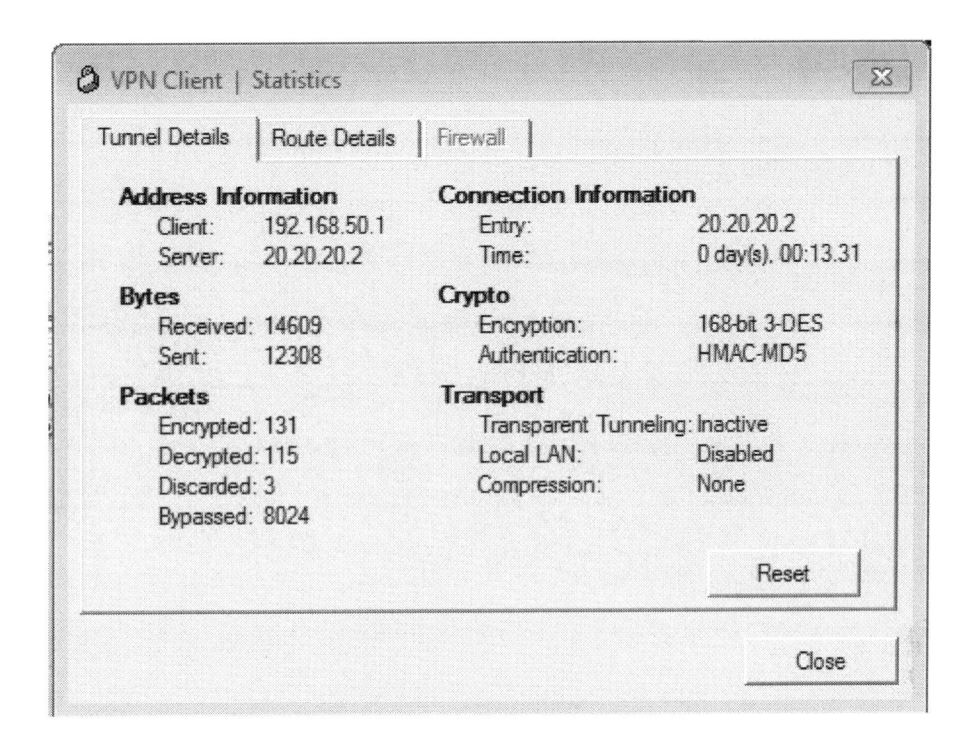

```
VPN Client | Statistics

Tunnel Details | Route Details | Firewall

Address Information              Connection Information
    Client:   192.168.50.1           Entry:    20.20.20.2
    Server:   20.20.20.2             Time:     0 day(s), 00:13.31
Bytes                            Crypto
    Received: 14609                  Encryption:      168-bit 3-DES
    Sent:     12308                  Authentication:  HMAC-MD5
Packets                          Transport
    Encrypted: 131                   Transparent Tunneling: Inactive
    Decrypted: 115                   Local LAN:             Disabled
    Discarded: 3                     Compression:           None
    Bypassed:  8024

                                                    Reset

                                                    Close
```

As you can see above, the remote user client received an IP address from the VPN pool (192.168.50.1) and is connected to the VPN server (Router R1) with IP 20.20.20.2. You can also see Encrypted and Decrypted packets which means that communication is working and it is secure.

You can find a complete configuration example of the scenario above in Chapter 4, Section 4.1.5.

2.1.4 Site-to-Site and Remote Access IPSEC VPN on same device

This is an example scenario that will combine all previous cases that we've described before. It's an excellent implementation to learn because it covers the majority of VPN connectivity options that an Enterprise requires. Specifically, we will cover static site-to-site IPSEC tunnel, dynamic site-to-site IPSEC tunnel and remote access tunnel, all on the same device. See diagram below:

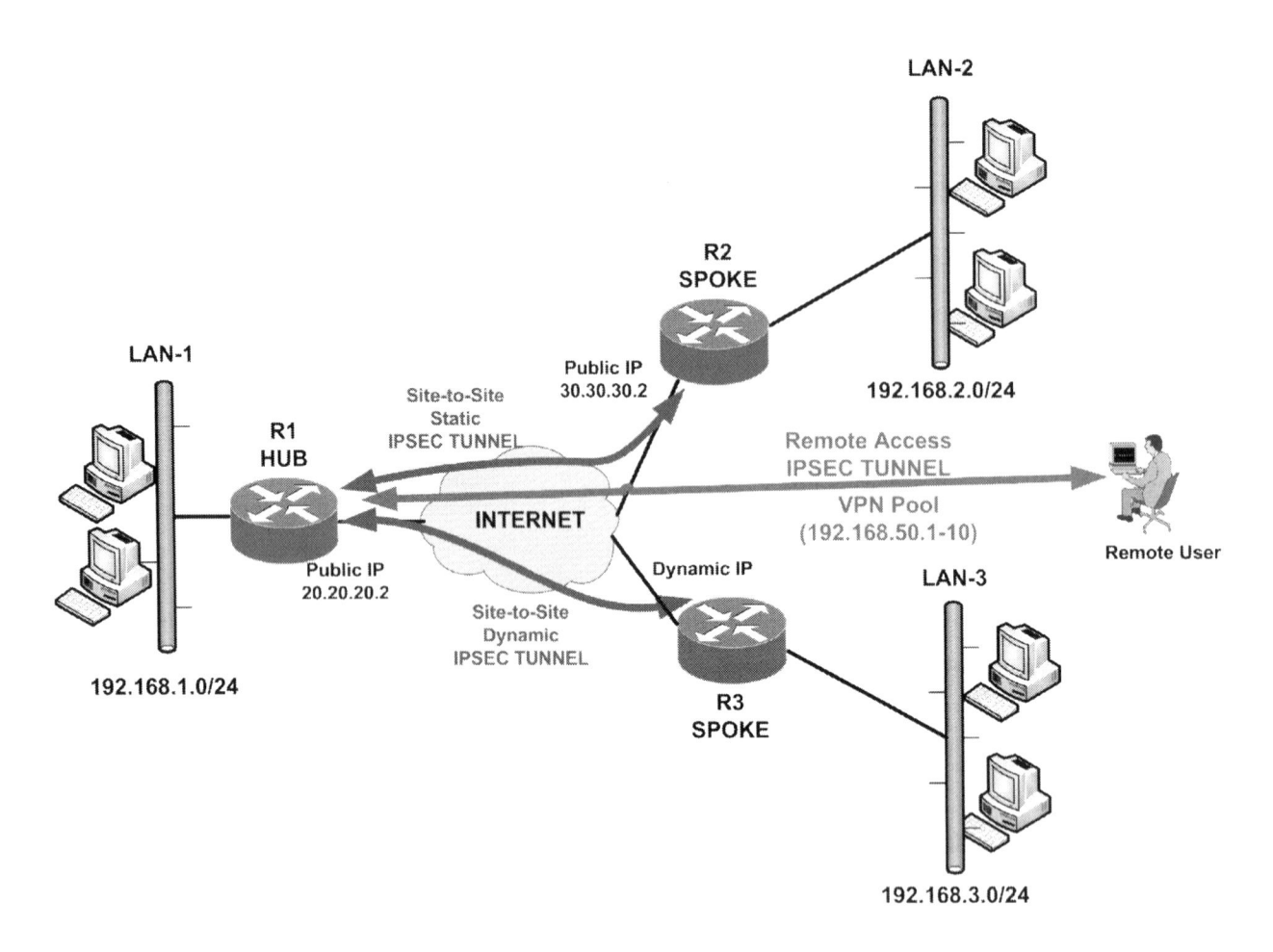

In our scenario, we have a Central Hub router (R1) with two remote branch offices (Spokes) which require site-to-site IPSEC VPN connectivity to the Central site. One of the remote branches (R2) has a static public IP whereas the second one (R3) has dynamic public IP. Also, the Central Hub Router will be configured to accept remote access VPN connections from distant users.

In our configuration we will utilize the newest IPSEC features, that is "**crypto keyring**" and "**isakmp profile**". These new features enable flexibility in IPSEC configuration and they are suitable in scenarios like the one we have here (that is, having a combination of VPN tunnels - such as static IP tunnel, dynamic IP tunnel and remote access tunnel - on the same device). Let's see the configuration below:

Router R1 (HUB):

- STEP 1: Configure Interesting Traffic and NAT Exemption

!First identify the Interesting traffic to be encrypted. We need to have 3 crypto ACLs, one for each remote site and one for the remote users pool.

R1(config)# ip access-list extended VPNsite1-ACL
R1(config-ext-nacl)# permit ip 192.168.1.0 0.0.0.255 192.168.2.0 0.0.0.255
R1(config)# ip access-list extended VPNsite2-ACL
R1(config-ext-nacl)# permit ip 192.168.1.0 0.0.0.255 192.168.3.0 0.0.0.255
R1(config)# ip access-list extended VPNclient-ACL
R1(config-ext-nacl)# permit ip 192.168.1.0 0.0.0.255 192.168.50.0 0.0.0.255

!Then exclude the VPN Interesting traffic from the NAT operation

R1(config)# ip access-list extended NAT-ACL
R1(config-ext-nacl)# deny ip 192.168.1.0 0.0.0.255 192.168.2.0 0.0.0.255
R1(config-ext-nacl)# deny ip 192.168.1.0 0.0.0.255 192.168.3.0 0.0.0.255
R1(config-ext-nacl)# deny ip 192.168.1.0 0.0.0.255 192.168.50.0 0.0.0.255
R1(config-ext-nacl)# permit ip 192.168.1.0 0.0.0.255 any

- STEP 2: Phase 1 IPSEC Configuration

!Now configure the Phase1 isakmp policy

R1(config)# crypto isakmp policy 10
R1(config-isakmp)# encryption 3des
R1(config-isakmp)# hash md5
R1(config-isakmp)# authentication pre-share
R1(config-isakmp)# group 2
R1(config-isakmp)# exit

Now we will see the **crypto keyring** feature. Each "keyring" is a repository of one or more pre-shared keys. The keyrings can be used later on in "isakmp profiles" as we will see.

R1(config)# crypto keyring vpnclientskey
R1(conf-keyring)# pre-shared-key address 0.0.0.0 0.0.0.0 key cisco123
R1(conf-keyring)# exit
R1(config)# crypto keyring staticbranch
R1(conf-keyring)# pre-shared-key address 30.30.30.2 key secretkey1
R1(conf-keyring)# exit
R1(config)# crypto keyring dynamicbranch
R1(conf-keyring)# pre-shared-key address 0.0.0.0 0.0.0.0 key secretkey2
R1(conf-keyring)# exit

As you can see above, we have configured 3 different keyrings, one for each type of VPN tunnel. So, we have a keyring for VPN remote access clients, for static IP remote branch and for dynamic IP remote branch. For example, if you had multiple static IP remote branches, you would configure their pre-shared key under the "staticbranch" keyring.

Moreover, the pre-shared keys which refer to dynamic IP remote nodes (e.g dynamic IP remote branch or remote VPN users), have as remote address "**0.0.0.0 0.0.0.0**" which means "**any IP**".

The keyrings above will be used in "**isakmp profiles**" as we will see next:

! Configuration of ISAKMP Profiles.

R1(config)# crypto isakmp profile staticL2L
R1(conf-isa-prof)# keyring staticbranch ←configured above
R1(conf-isa-prof)# match identity address 30.30.30.2 255.255.255.255 ←Router R2 static IP
R1(conf-isa-prof)# exit

R1(config)# **crypto isakmp profile** dynamicL2L
R1(conf-isa-prof)# **keyring** dynamicbranch ←configured above
R1(conf-isa-prof)# **match identity address 0.0.0.0** ←Any remote branch with dynamic IP can connect
R1(config)# **crypto isakmp profile** remoteclients
R1(conf-isa-prof)# **keyring** vpnclientskey ←configured above
R1(conf-isa-prof)# **match identity group remotevpn** ←See section 2.1.3 above
R1(conf-isa-prof)# **client authentication list userauthen** ←See section 2.1.3 above
R1(conf-isa-prof)# **isakmp authorization list groupauthor** ←See section 2.1.3 above
R1(conf-isa-prof)# **client configuration address respond** ←See section 2.1.3 above

NOTE:
For remote VPN clients, you need to configure also "isakmp client configuration group", username, and authentication lists as described in section 2.1.3 above.

The "isakmp profiles" configured above will be used in "crypto maps" as we will see next:

- **STEP 3: Phase 2 IPSEC Configuration**

!Phase2 Transform Set and Crypto Map configuration

R1(config)# crypto ipsec transform-set TRSET esp-3des esp-md5-hmac
R1(cfg-crypto-trans)# exit

! Now we must configure both a dynamic and a static crypto map to accommodate all possible
!VPN tunnel types we have. The dynamic crypto map will have two entries (one for the dynamic
!branch and one for the remote VPN clients).

R1(config)# crypto dynamic-map DYNMAP 10
R1(config-crypto-map)# set transform-set TRSET
R1(config-crypto-map)# **set isakmp-profile** remoteclients ←Attach isakmp profile of VPN clients configured before
R1(config-crypto-map)# exit

R1(config)# crypto dynamic-map DYNMAP 20

R1(config-crypto-map)# set transform-set TRSET

R1(config-crypto-map)# set isakmp-profile dynamicL2L ←Attach isakmp profile of dynamic IP branch configured before

R1(config-crypto-map)# match address VPNsite2-ACL

R1(config-crypto-map)# exit

R1(config)# crypto map VPNMAP 10 ipsec-isakmp ←This is the static tunnel crypto map

R1(config-crypto-map)# set peer 30.30.30.2

R1(config-crypto-map)# set transform-set TRSET

R1(config-crypto-map)# set isakmp-profile staticL2L ←Attach isakmp profile of static IP branch configured before

R1(config-crypto-map)# match address VPNsite1-ACL

R1(config-crypto-map)# exit

R1(config)# crypto map VPNMAP 20 ipsec-isakmp dynamic DYNMAP ←Attach the dynamic crypto map on a new entry of the static crypto map.

!Attach the static crypto map VPNMAP to the WAN outside interface (FE0/1)of the Hub router

R1(config)# interface FastEthernet0/1

R1(config-if)# crypto map VPNMAP

Important Points to Remember:

- Create a keyring for each Type of VPN tunnel (static IP branch, dynamic IP branch, VPN clients).
- Attach each keyring to a corresponding "isakmp profile".
- Create one dynamic crypto map with two entries. On one entry attach the "isakmp profile" for the dynamic Lan-to-Lan tunnel and on the other entry attach the "isakmp profile" for the remote clients tunnel.
- Create one static crypto map with two entries. On one entry attach the "isakmp profile" of the static Lan-to-Lan tunnel, and on the second entry attach the dynamic crypto map.

You can find a complete configuration example of the scenario above (which includes also configuration of the remote branch routers) in Chapter 4, Section 4.1.6.

2.2 Route-Based VPN Configuration on Cisco Routers

On section 2.1 we've seen Policy-Based VPN implementation, which is basically the traditional IPSEC VPN configuration. In this section we will see a more flexible VPN type, Route-Based VPN.

2.2.1 Site-to-Site VPN Using GRE with IPSEC Protection

This is the simplest scenario of having two sites connected over the Internet with GRE tunnel protected by IPSEC. The IPSEC protection provides security and it's highly recommended, but it's not a requirement for having network connectivity between the two sites. You can have only GRE tunnel if you want without IPSEC. The major advantage of using GRE is that it can transport multicast traffic and dynamic routing protocols between the two sites.

Let's see now step by step how to configure this topology:

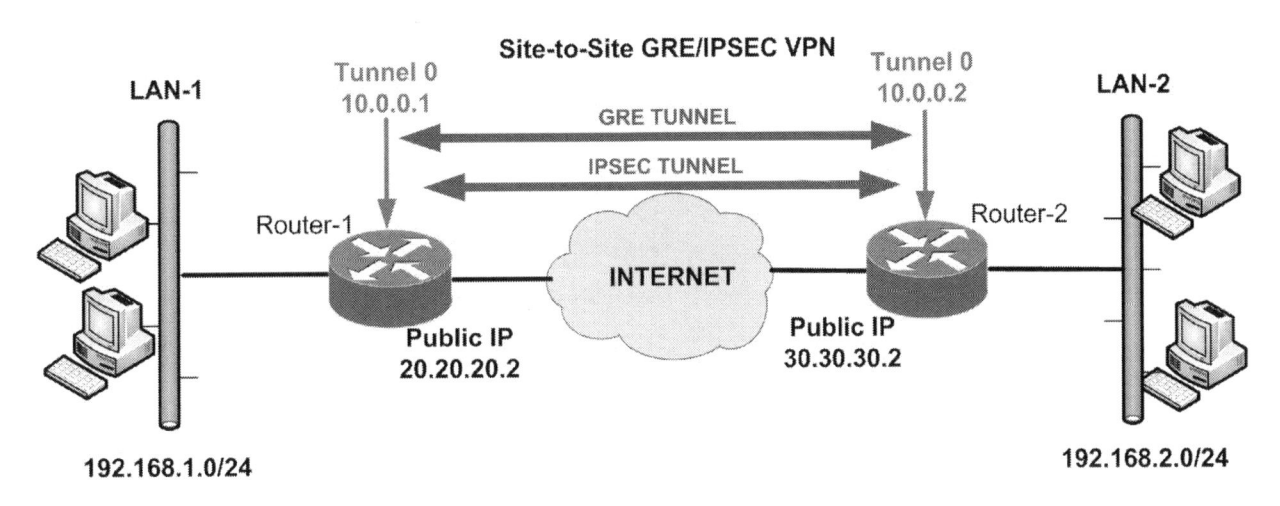

- **STEP 1: Configure the GRE Tunnel**

First let's setup the GRE tunnel between the two sites. A GRE VPN tunnel is nothing more than a Virtual Tunnel Interface with the appropriate settings. All traffic passing through this Tunnel is encapsulated in GRE.

Router-1:

R1(config)# interface Tunnel0 ←Virtual Tunnel
R1(config-if)# ip address 10.0.0.1 255.255.255.0 ←Choose a private IP for the Tunnel
R1(config-if)# tunnel source 20.20.20.2 ←This is the public IP of WAN interface
R1(config-if)# tunnel destination 30.30.30.2 ←This is the public IP of remote router
R1(config-if)# exit

Router-2:

R2(config)# interface Tunnel0
R2(config-if)# ip address 10.0.0.2 255.255.255.0
R2(config-if)# tunnel source 30.30.30.2
R2(config-if)# tunnel destination 20.20.20.2
R2(config-if)# exit

- **STEP 2: Configure Routing to reach the remote subnets**

In this example we will use static routing to reach remote networks. You can use also dynamic routing protocols as we will see in other examples.

Router-1:

R1(config)# ip route 192.168.2.0 255.255.255.0 10.0.0.2←LAN-2 subnet can be reached via the other end of GRE Tunnel (10.0.0.2)

Router-2:

R2(config)# ip route 192.168.1.0 255.255.255.0 10.0.0.1←LAN-1 subnet can be reached via the other end of GRE Tunnel (10.0.0.1)

This concludes the configuration of the GRE Tunnel. Now, LAN-1 and LAN-2 private networks can communicate between each other over the Internet. However, we recommend encrypting and securing this GRE tunnel with IPSEC as we will see next:

- **STEP 3: Configure IPSEC Phase 1**

Assuming that we want IPSEC protection of the GRE tunnel (which is recommended), we still have to configure IPSEC Phase 1 and Phase 2 similar to the previous "Policy-Based" VPN configurations.

Router-1:

R1(config)# crypto isakmp policy 10
R1(config-isakmp)# encryption 3des
R1(config-isakmp)# hash md5
R1(config-isakmp)# authentication pre-share
R1(config-isakmp)# group 2
R1(config-isakmp)# exit

R1(config)# crypto isakmp key testkey123 address 30.30.30.2

Router-2:

R2(config)# crypto isakmp policy 10
R2(config-isakmp)# encryption 3des
R2(config-isakmp)# hash md5
R2(config-isakmp)# authentication pre-share
R2(config-isakmp)# group 2
R2(config-isakmp)# exit

R2(config)# crypto isakmp key testkey123 address 20.20.20.2

- **STEP 4: Configure IPSEC Phase 2 (ipsec profile)**

As you remember from previous configurations, in Phase 2 of IPSEC we had to define a "Transform Set" and "Crypto Map". However, a "Crypto Map" is not needed in our case. We will introduce a new feature, called "**ipsec profile**" instead of "crypto map". The "**ipsec profile**" will be attached to the GRE Tunnel interface to enable IPSEC protection of the GRE.

Router-1:

R1(config)# crypto ipsec transform-set TRSET esp-3des esp-md5-hmac ←This is the transform set which defines the encryption and authentication protocols of IPSEC
R1(cfg-crypto-trans)# exit

R1(config)# crypto ipsec profile GRE-PROTECTION←This is the ipsec profile that will be used to protect the GRE tunnel.
R1(ipsec-profile)# set transform-set TRSET ← Attach the transform set configured above
R1(ipsec-profile)# exit

Router-2:

R2(config)# crypto ipsec transform-set TRSET esp-3des esp-md5-hmac ←This is the transform set which defines the encryption and authentication protocols of IPSEC
R2(cfg-crypto-trans)# exit

R2(config)# crypto ipsec profile GRE-PROTECTION←This is the ipsec profile that will be used to protect the GRE tunnel.
R2(ipsec-profile)# set transform-set TRSET ← Attach the transform set configured above
R2(ipsec-profile)# exit

- **STEP 5: Attach the IPSEC Protection profile to GRE Tunnel**

Router-1:

R1(config)# interface Tunnel0
R1(config-if)# tunnel protection ipsec profile GRE-PROTECTION←Attach the IPSEC Profile configured before in order to provide security to GRE Tunnel.

Router-2:

R2(config)# interface Tunnel0

R2(config-if)# tunnel protection ipsec profile GRE-PROTECTION ← Attach the IPSEC Profile configured before in order to provide security to GRE Tunnel.

You can find a complete configuration example of the scenario above in Chapter 4, Section 4.1.7.

Important Points to Remember:

- As you have seen above, in Route-Based VPNs there is no need to define Interesting Traffic using ACLs as we did in Policy-Based VPNs.
- Instead of defining which Interesting Traffic will be encrypted, configure static (or dynamic) routing to tell the router that the remote LAN network will be reachable via the other-end of the GRE Tunnel Interface. All traffic routed through the Tunnel Interface will be placed in the VPN and also protected by IPSEC. That's why we call this VPN type as "Route-Based" VPN.
- There is no need to exclude Interesting Traffic from NAT operation.
- If IPSEC protection is required, just configure an "**isakmp policy**", "**pre-shared key**", "**transform-set**" and "**ipsec profile**". No "**crypto map**" is required.
- After configuring the required settings for IPSEC, just attach the "**ipsec profile**" to the GRE Tunnel Interface.

2.2.2 Hub-and-Spoke VPN Using GRE with IPSEC Protection

This configuration example is the same as the site-to-site GRE VPN as far as configuration on the Spoke sites is concerned. However, the HUB router is configured with additional GRE Tunnel Interfaces in order to create a point-to-point GRE tunnel with each remote branch router. This makes the GRE VPN not very scalable with large networks. Also, GRE VPN topologies require all sites to have static public IP which is another disadvantage.

Also, in this scenario it's a good idea to implement a dynamic routing protocol (EIGRP in our case) between the three sites in order to advertise network routes dynamically over the GRE/IPSEC tunnels.

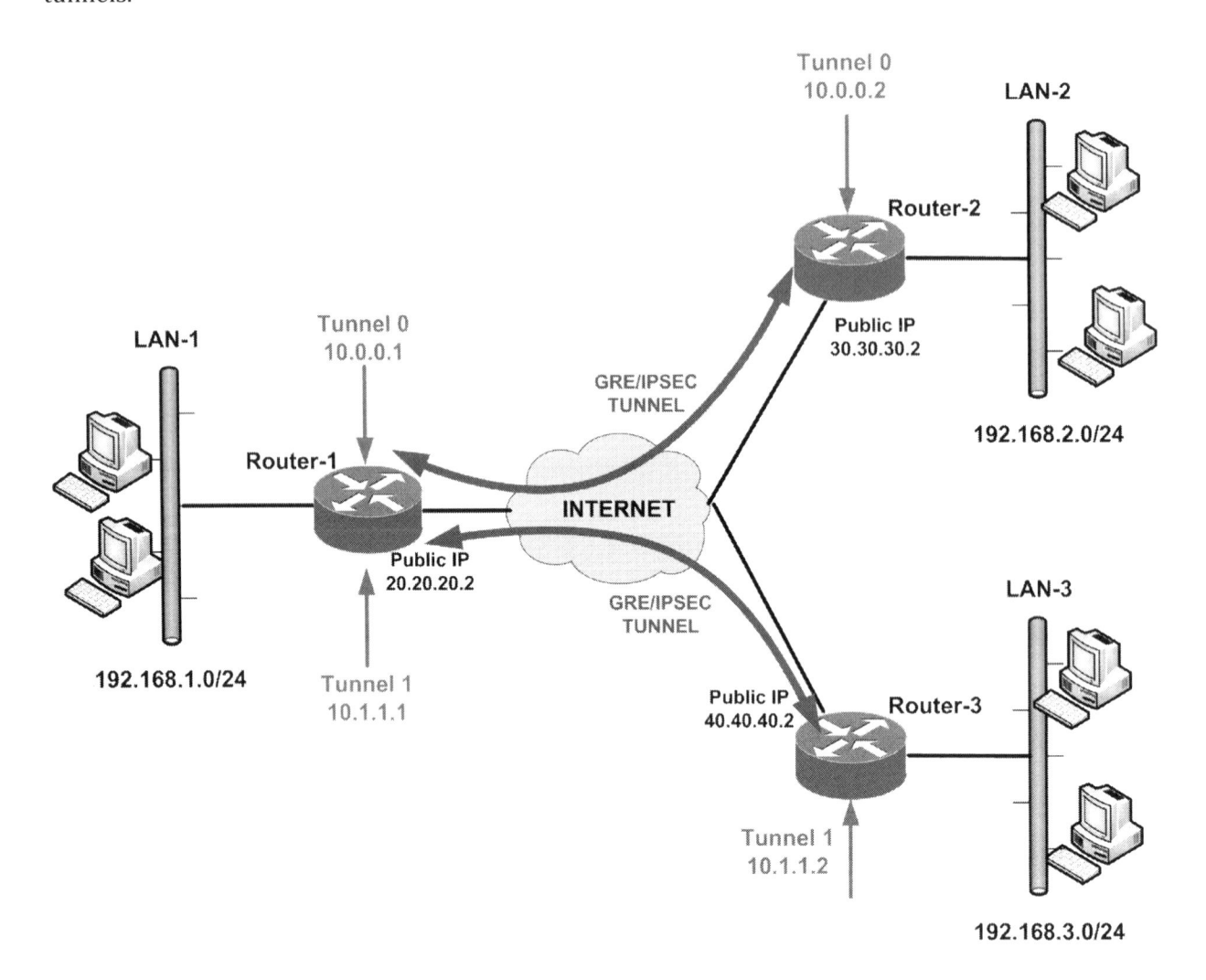

As shown from the network above, we have two point to point GRE/IPSEC tunnels between the Hub Router (Router-1) and the Spoke Routers (Router-2, Router-3). Each point-to-point GRE Tunnel must have its own private network subnet (for example 10.0.0.0/30 and 10.1.1.0/30).

We will describe the configuration of the Hub Router only since Spoke Routers are configured the same as site-to-site GRE VPN we've seen before.

- **STEP 1: Configure the GRE Tunnels**

Router-1:

R1(config)# interface Tunnel0 ←GRE Tunnel for site-to-site VPN with Router-2
R1(config-if)# ip address 10.0.0.1 255.255.255.0 ←Choose a private IP for the Tunnel
R1(config-if)# tunnel source 20.20.20.2 ←This is the public IP of WAN interface
R1(config-if)# tunnel destination 30.30.30.2 ←This is the public IP of remote Router-2
R1(config-if)# exit
R1(config)# interface Tunnel1 ←GRE Tunnel for site-to-site VPN with Router-3
R1(config-if)# ip address 10.1.1.1 255.255.255.0 ←Choose a private IP for the Tunnel
R1(config-if)# tunnel source 20.20.20.2 ←This is the public IP of WAN interface
R1(config-if)# tunnel destination 40.40.40.2 ←This is the public IP of remote Router-3
R1(config-if)# exit

- **STEP 2: Configure Routing to reach the remote subnets**

After setting up the two point-to-point GRE Tunnels, we need to configure routing so that devices will advertise their local LAN and Tunnel IP addresses to the other routers participating in the GRE VPN topology. In this particular example we will use EIGRP. You can use also static routes (if you don't have many sites) or OSPF.

Router-1:

R1(config)# router eigrp 100←Create an EIGRP instance with Autonomous System 100
R1(config-router)# no auto-summary←Disable automatic summarization of subnets
R1(config-router)# network 10.0.0.0 0.0.0.255←Tunnel0 subnet
R1(config-router)# network 10.1.1.0 0.0.0.255←Tunnel1 subnet
R1(config-router)# network 192.168.1.0 0.0.0.255←LAN-1 subnet
R1(config-router)# exit

When configuring EIGRP, you specify which router networks are included in EIGRP using the **network** command. The router will send EIGRP messages and try to establish adjacencies with other EIGRP speaking routers off these network interfaces. By including the Tunnel Interface networks, this means that routers will try to establish EIGRP adjacencies with other routers via the GRE VPN Tunnel Interfaces.

Let's see the other two routers' configuration regarding EIGRP:

Router-2:

R2(config)# **router eigrp 100** ←Create an EIGRP instance with Autonomous System 100

R2(config-router)# **no auto-summary** ←Disable automatic summarization of subnets

R2(config-router)# **network 10.0.0.0 0.0.0.255** ←Tunnel0 subnet

R2(config-router)# **network 192.168.2.0 0.0.0.255** ←LAN-2 subnet

R2(config-router)# **exit**

Router-3:

R3(config)# **router eigrp 100** ←Create an EIGRP instance with Autonomous System 100

R3(config-router)# **no auto-summary** ←Disable automatic summarization of subnets

R3(config-router)# **network 10.1.1.0 0.0.0.255** ←Tunnel1 subnet

R3(config-router)# **network 192.168.3.0 0.0.0.255** ←LAN-3 subnet

R3(config-router)# **exit**

Let's see the routing Tables of the Hub and one of the Spoke routers:

Router-1 (HUB):

```
R1# show ip route

….(Output Omitted)
C    10.1.1.0 is directly connected, Tunnel1
C    10.0.0.0 is directly connected, Tunnel0
C    192.168.1.0/24 is directly connected, FastEthernet0/1
D    192.168.2.0/24 [90/297270016] via 10.0.0.2, 00:13:22, Tunnel0
D    192.168.3.0/24 [90/297270016] via 10.1.1.2, 00:09:25, Tunnel1
….(Output Omitted)
```

As shown above, the Hub router learns the remote LAN networks (192.168.2.0/24 and 192.168.3.0/24) from EIGRP (Denoted as "**D**") via the GRE Tunnel Interfaces.

Router-2 (Spoke):

```
R2# show ip route
….(Output Omitted)
D    10.1.1.0 [90/310044416] via 10.0.0.1, 00:15:02, Tunnel0
C    10.0.0.0 is directly connected, Tunnel0
D    192.168.1.0/24 [90/297270016] via 10.0.0.1, 00:15:02, Tunnel0
C    192.168.2.0/24 is directly connected, FastEthernet0/1
D    192.168.3.0/24 [90/310070016] via 10.0.0.1, 00:11:06, Tunnel0
….(Output Omitted)
```

As shown above, one of the Spoke routers learns the remote LAN networks (192.168.1.0/24 and 192.168.3.0/24) from EIGRP (Denoted as "**D**") via the GRE Tunnel Interface (Tunnel0). Also, this Spoke router can reach the other Spoke's LAN (192.168.3.0/24) via the HUB (Tunnel0 Interface). By running therefore a dynamic routing protocol in a GRE Hub-and-Spoke topology, the spoke sites can reach each other via the Hub site.

- **STEP 3: Configure IPSEC Phase 1**

Router-1:

R1(config)# crypto isakmp policy 10
R1(config-isakmp)# encryption 3des
R1(config-isakmp)# hash md5
R1(config-isakmp)# authentication pre-share
R1(config-isakmp)# group 2
R1(config-isakmp)# exit

R1(config)# crypto isakmp key testkey123 address 30.30.30.2
R1(config)# crypto isakmp key testkey1234 address 40.40.40.2

As you can see above, we have two pre-shared keys, one for each Spoke branch router.

- STEP 4: Configure IPSEC Phase 2 (ipsec profile)

Router-1:

R1(config)# crypto ipsec transform-set TRSET esp-3des esp-md5-hmac

R1(cfg-crypto-trans)# exit

R1(config)# crypto ipsec profile GRE-PROTECTION

R1(ipsec-profile)# set transform-set TRSET

R1(ipsec-profile)# exit

- STEP 5: Attach the IPSEC Protection profile to GRE Tunnels

Router-1:

R1(config)# interface Tunnel0

R1(config-if)# tunnel protection ipsec profile GRE-PROTECTION

R1(config)# interface Tunnel1

R1(config-if)# tunnel protection ipsec profile GRE-PROTECTION

You can find a complete configuration example of the scenario above in Chapter 4, Section 4.1.8.

2.2.3 VPN Using Static Virtual Tunnel Interface (SVTI)

The configuration for this VPN type is exactly the same as the site-to-site GRE VPN (2.2.1) except to one small difference. Under the Tunnel Interface you need to specify the mode as "**tunnel mode ipsec ipv4**". This mode denotes that the Tunnel is VTI. If you don't specify a "tunnel mode" for the Tunnel, it takes the default mode which is GRE.

![Network diagram showing Site-to-Site IPSEC VPN between Router-1 (Public IP 20.20.20.2, Tunnel 0 10.0.0.1, tunnel mode ipsec ipv4) on LAN-1 (192.168.1.0/24) and Router-2 (Public IP 30.30.30.2, Tunnel 0 10.0.0.2, tunnel mode ipsec ipv4) on LAN-2 (192.168.2.0/24) connected via the Internet with an IPSEC VTI TUNNEL.]

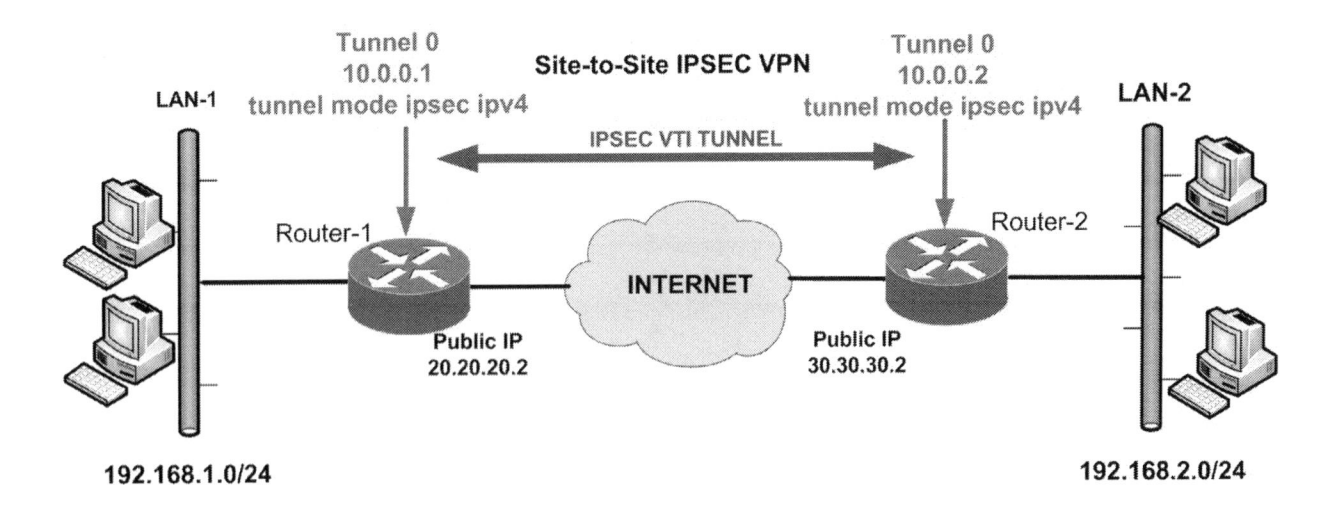

We won't see the configuration of the above in detail since it's the same as section 2.2.1. Just keep in mind the Tunnel mode configuration:

Router-1:

R1(config)# interface Tunnel0 ←Static VTI Tunnel for site-to-site VPN with Router-2

R1(config-if)# ip address 10.0.0.1 255.255.255.0 ←Choose a private IP for the Tunnel

R1(config-if)# tunnel source 20.20.20.2 ←This is the public IP of WAN interface

R1(config-if)# tunnel destination 30.30.30.2 ←This is the public IP of remote Router-2

R1(config-if)# tunnel mode ipsec ipv4 ←Configure the Tunnel as VTI and not GRE

R1(config-if)# tunnel protection ipsec profile VTI-PROTECTION ←Attach an IPSEC profile for protection

R1(config-if)# exit

2.2.4 VPN Using Dynamic Virtual Tunnel Interface (DVTI)

This VPN Type is suitable for creating scalable Hub-and-Spoke topologies which can accommodate a large number of remote branch offices. The HUB router will use Dynamic VTI while the Spoke routers will be configured as Static VTI, as we've described in Chapter 1 above. The Spoke Branch routers can have dynamic WAN Public IP. Here we will see the configuration details of this VPN type. Specifically, we will describe the HUB and one Spoke router.

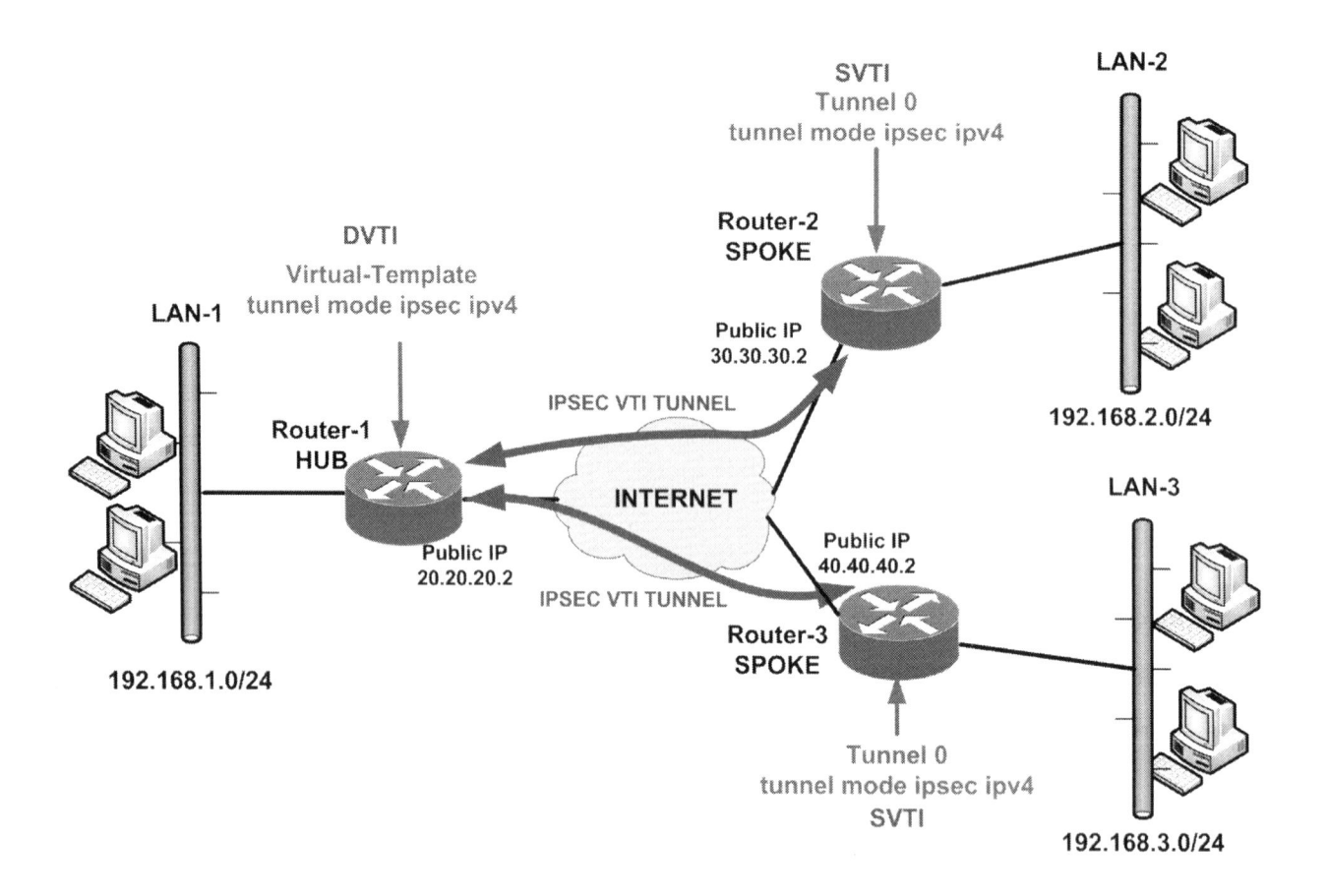

- **STEP 1: Configure the DVTI and SVTI Tunnels**

The HUB Router will be configured with Dynamic VTI whereas the Spoke Routers with Static VTIs.

Router-1 (HUB):
DVTI Configuration

Instead of having an "Interface Tunnel" (as used in Static VTI or GRE), for DVTI we need to configure a single "**Virtual Template**" Interface which will accommodate all Spoke sites. When a Spoke site connects to the Hub router, a "**Virtual-Access**" logical interface is created for each Spoke from the "Virtual Template" Interface. Therefore, each remote Spoke Branch router will have its own "Virtual-Access" logical interface on the HUB. These interfaces can be seen by running the "**show ip interface brief**" command on the HUB.

R1(config)# interface FastEthernet0/0 ←Public WAN Interface
R1(config-if)# ip address 20.20.20.2 255.255.255.0 ←Public IP
R1(config-if)# exit

R1(config)# interface Loopback0 ←To be used for Virtual-Template below
R1(config-if)# ip address 10.0.0.1 255.255.255.0←Choose an unused private IP
R1(config-if)# exit

R1(config-if)# interface Virtual-Template1 type tunnel←This is the DVTI Interface
R1(config-if)# ip unnumbered Loopback0←Configured above
R1(config-if)# tunnel source FastEthernet0/0←Source of DVTI is the WAN Interface
R1(config-if)# tunnel mode ipsec ipv4←VTI mode
R1(config-if)# exit

Router-2 (SPOKE):

Spoke routers are configured with normal Static VTI using Tunnel Interfaces as we've described on previous sections.

R2(config)# interface FastEthernet0/0 ←Public WAN Interface
R2(config-if)# ip address 30.30.30.2 255.255.255.0 ←Public IP (can be dynamic also)
R2(config-if)# exit
R2(config)# interface Loopback0 ←To be used for Tunnel Interface below
R2(config-if)# ip address 10.1.1.1 255.255.255.0 ←Choose an unused private IP
R2(config-if)# exit

R2(config)# interface Tunnel0 ←This is the SVTI Interface
R2(config-if)# ip unnumbered Loopback0 ←Configured above
R2(config-if)# tunnel source FastEthernet0/0 ←Source of SVTI is the WAN Interface
R2(config-if)# tunnel destination 20.20.20.2 ←WAN IP of remote HUB Router
R2(config-if)# tunnel mode ipsec ipv4 ←VTI mode
R2(config-if)# exit

- **STEP 2: Configure Routing to reach the remote subnets**

After setting up the VTI Tunnels, we need to configure routing so that devices will advertise their local LAN and Tunnel IP addresses to the other routers participating in the VPN topology. In this particular example we will use EIGRP.

Router-1:

R1(config)# router eigrp 100 ←Create an EIGRP instance with Autonomous System 100
R1(config-router)# no auto-summary ←Disable automatic summarization of subnets
R1(config-router)# network 10.0.0.0 0.0.0.255 ←Subnet of Virtual-Template private IP
R1(config-router)# network 192.168.1.0 0.0.0.255 ←LAN-1 subnet
R1(config-router)# exit

Router-2:

R2(config)# router eigrp 100 ←Create an EIGRP instance with Autonomous System 100
R2(config-router)# no auto-summary ←Disable automatic summarization of subnets
R2(config-router)# network 10.1.1.0 0.0.0.255 ←Subnet of Tunnel private IP
R2(config-router)# network 192.168.2.0 0.0.0.255 ←LAN-2 subnet
R2(config-router)# exit

Let's see the routing Tables and EIGRP Neighbors of the Hub and Spoke routers:

Router-1 (HUB):

```
R1# show ip interface brief

Interface          IP-Address       OK?   Method   Status   Protocol
FastEthernet0/0    20.20.20.2       YES   manual   up       up
FastEthernet0/1    192.168.1.1      YES   manual   up       up
NVI0               unassigned       NO    unset    up       up
Virtual-Access1    unassigned       YES   unset    down     down
Virtual-Template1  10.0.0.1         YES   TFTP     down     down
Virtual-Access2    10.0.0.1         YES   TFTP     up       up
Virtual-Access3    10.0.0.1         YES   TFTP     up       up
Loopback0          10.0.0.1         YES   manual   up       up
```

As described before, "Virtual-Access" Interfaces are created (see *Virtual-Access2* and *Virtual-Access3* above) from the Virtual-Template interface for each Spoke router.

```
R1# show ip eigrp neighbors

IP-EIGRP neighbors for process 100
H   Address         Interface   Hold   Uptime   SRTT  RTO  Q  Seq
                                (sec)  (ms)              Cnt Num
1   10.1.1.1        Vi2         10     00:21:16 128  5000 0  23
0   10.2.2.1        Vi3         11     00:22:09 153  5000 0  3
```

Router R1 (Hub) established two EIGRP neighbors with the two Spoke sites via the Virtual-Access interfaces.

```
R1# show ip route

....(Output Omitted)
     20.0.0.0/24 is subnetted, 1 subnets
C        20.20.20.0 is directly connected, FastEthernet0/0
     10.0.0.0/24 is subnetted, 3 subnets
D        10.2.2.0 [90/297372416] via 10.2.2.1, 00:22:31, Virtual-Access3
D        10.1.1.0 [90/297372416] via 10.1.1.1, 00:21:39, Virtual-Access2
C        10.0.0.0 is directly connected, Loopback0
C    192.168.1.0/24 is directly connected, FastEthernet0/1
D    192.168.2.0/24 [90/297246976] via 10.1.1.1, 00:21:39, Virtual-Access2
D    192.168.3.0/24 [90/297246976] via 10.2.2.1, 00:22:32, Virtual-Access3
....(Output Omitted)
```

The routing Table of R1 is filled with the networks of the Spoke Branch sites as shown above.

<u>Router-2 (SPOKE):</u>

```
R2# show ip eigrp neighbors

IP-EIGRP neighbors for process 100
H   Address         Interface   Hold   Uptime   SRTT  RTO  Q  Seq
                                (sec)  (ms)              Cnt Num
0   10.0.0.1        Tu0         11     00:24:20 154  5000 0  17
```

```
R2# show ip route

...(Output Omitted)

    10.0.0.0/24 is subnetted, 3 subnets
D    10.2.2.0 [90/310172416] via 10.0.0.1, 00:24:47, Tunnel0
C    10.1.1.0 is directly connected, Loopback0
D    10.0.0.0 [90/297372416] via 10.0.0.1, 00:24:47, Tunnel0
D   192.168.1.0/24 [90/297270016] via 10.0.0.1, 00:24:47, Tunnel0
C   192.168.2.0/24 is directly connected, FastEthernet0/1
D   192.168.3.0/24 [90/310046976] via 10.0.0.1, 00:24:47, Tunnel0
    30.0.0.0/24 is subnetted, 1 subnets
C    30.30.30.0 is directly connected, FastEthernet0/0

...(Output Omitted)
```

The Spoke router learns all required networks from the Hub Router (via Tunnel0).

- **STEP 3: Configure IPSEC Phase 1**

The Hub router will accept connections from any remote IP address. Also, we must use "keyring" and "isakmp profile" for the Hub router in order to bind the "Virtual-Template" interface to IPSEC Phase 1.

Router-1 (HUB):

R1(config)# crypto keyring remotebranchkeys
R1(conf-keyring)# pre-shared-key address 0.0.0.0 0.0.0.0 key strongkey123
R1(conf-keyring)# exit

R1(config)# crypto isakmp policy 10
R1(config-isakmp)# encryption 3des
R1(config-isakmp)# hash md5
R1(config-isakmp)# authentication pre-share
R1(config-isakmp)# group 2
R1(config-isakmp)# exit

R1(config)# crypto isakmp profile DVTI ←Isakmp Profile
R1(conf-isa-prof)# keyring remotebranchkeys
R1(conf-isa-prof)# match identity address 0.0.0.0
R1(conf-isa-prof)# virtual-template 1 ←Attach Virtual-Template1 configured before
R1(conf-isa-prof)# exit

Router-2 (SPOKE):

For the Spoke we don't need "keyring" and "isakmp profile" configuration. The traditional IPSEC Phase 1 configuration is enough.

R2(config)# crypto isakmp policy 10
R2(config-isakmp)# encryption 3des
R2(config-isakmp)# hash md5
R2(config-isakmp)# authentication pre-share
R2(config-isakmp)# group 2
R2(config-isakmp)# exit

R2(config)# crypto isakmp key strongkey123 address 20.20.20.2

- STEP 4: Configure IPSEC Phase 2 (ipsec profile)

Router-1 (HUB):

R1(config)# crypto ipsec transform-set TRSET esp-3des esp-md5-hmac
R1(cfg-crypto-trans)# exit

R1(config)# crypto ipsec profile VTI-PROTECTION
R1(ipsec-profile)# set transform-set TRSET
R1(ipsec-profile)# exit

Router-2 (SPOKE):

R2(config)# crypto ipsec transform-set TRSET esp-3des esp-md5-hmac

R2(cfg-crypto-trans)# exit

R2(config)# crypto ipsec profile VTI-PROTECTION

R2(ipsec-profile)# set transform-set TRSET

R2(ipsec-profile)# exit

- STEP 5: Attach the IPSEC Protection profile to VTI Tunnels

Router-1 (HUB):

R1(config)# interface Virtual-Template1 type tunnel

R1(config-if)# tunnel protection ipsec profile VTI-PROTECTION

Router-2 (SPOKE):

R2(config)# interface Tunnel0

R2(config-if)# tunnel protection ipsec profile VTI-PROTECTION

You can find a complete configuration example of the scenario above in Chapter 4, Section 4.1.9.

2.3 Dynamic Multipoint VPN (DMVPN)

DMVPN provides full meshed connectivity for large Hub-and-Spoke network topologies. The first-time configuration is a little bit complex, but after you get it working you can start adding Spoke sites without changing the Hub configuration. Also, spoke sites with dynamic WAN IP are supported as well. Moreover, there is no need to manually configure Spoke-to-Spoke VPN tunnels since these are created automatically whenever there is traffic between spokes.

In our network scenario below, we have one central Hub router with two remote Spoke branch sites (one with static public IP and one with Dynamic public IP). In our configuration we will see new features and new protocols that we haven't seen before. These include **Multipoint GRE** and **Next Hop Resolution Protocol (NHRP)**. In Chapter 1, Section 1.4 above we described how DMVPN works in general. Here we will get into the actual configuration. We will see how to configure the Hub (Router1) and one of the Spoke sites (Router3). All spoke sites will be similar in configuration.

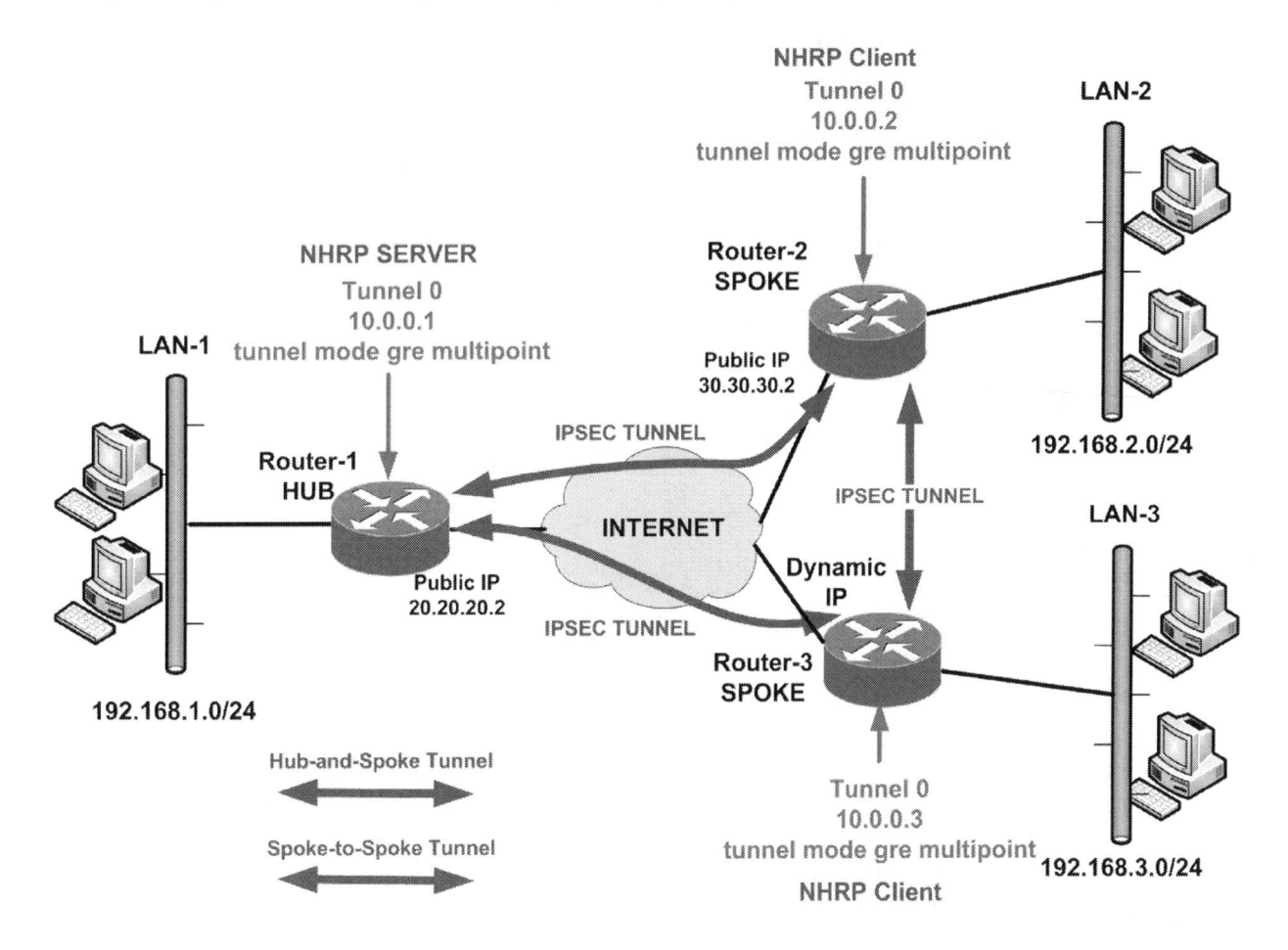

- **STEP 1: Configure the Multipoint GRE Tunnels**

All routers in the topology (Hub and Spoke Routers) will be configured with Multipoint GRE (mGRE) Tunnels. An mGRE tunnel is a single Tunnel Interface that can have multiple GRE connections with other routers. In previous VPN types (point-to-point GRE and static VTI) we had a dedicated Tunnel Interface for each point-to-point VPN tunnel. Now, using mGRE we can have a single Tunnel Interface for multiple VPN tunnel connections. This means that we don't need to have a separate Tunnel Interface on the Hub for each new Spoke added in the topology, thus making DMVPN very scalable.

Router-1 (HUB):
mGRE Configuration

For mGRE Tunnel, we configure the mode of Tunnel Interface as "**tunnel mode gre multipoint**". Also, unlike the GRE and VTI tunnel interfaces we've seen before, there is no need to define a "**tunnel destination**" command. This means that you don't need to have a defined static IP for remote sites, which allows for dynamic IP branch sites to participate in the DMVPN.

R1(config)# interface FastEthernet0/1 ←Public WAN Interface on Hub Router
R1(config-if)# ip address 20.20.20.2 255.255.255.0 ←Public IP
R1(config-if)# exit

R1(config)# interface Tunnel0 ←This is the mGRE Interface
R1(config-if)# ip address 10.0.0.1 255.255.255.0 ←Select a private IP subnet
R1(config-if)# tunnel source FastEthernet0/1 ←Source of Tunnel is the WAN interface
R1(config-if)# tunnel mode gre multipoint ←Set the tunnel mode as mGRE
R1(config-if)# ip mtu 1440 ←Reduce the MTU to allow extra overhead from mGRE and IPSEC

Router-3 (SPOKE):

R3(config)# interface FastEthernet0/1 ←Public WAN Interface on Spoke Router
R3(config-if)# ip address dhcp ←Dynamic Public IP via DHCP
R3(config-if)# exit

R3(config)# interface Tunnel0 ←This is the mGRE Interface
R3(config-if)# ip address 10.0.0.3 255.255.255.0 ←IP of Tunnel in the same subnet as Hub
R3(config-if)# tunnel source FastEthernet0/1 ←Source of Tunnel is the WAN interface
R3(config-if)# tunnel mode gre multipoint ←Set the tunnel mode as mGRE
R3(config-if)# ip mtu 1440 ←Reduce the MTU to allow extra overhead from mGRE and IPSEC

- **STEP 2: Configure Next Hop Resolution Protocol (NHRP)**

Router-1 (HUB):

NHRP Configuration

NHRP is used to provide IP mapping between WAN IP address and Tunnel IP address. The Hub is the NHRP Server (NHS) while the Spoke sites are the NHRP Clients. The clients (Spoke sites) query the NHS router (Hub site) to obtain the physical WAN public IP of other Spoke routers. This helps the Spoke routers to dynamically establish VPN tunnels with other Spoke routers in the topology. It doesn't matter if a Spoke router has dynamic public IP. As long as this Spoke router will register its current public IP with the NHS server, all the other Spoke sites will find it and will be able to establish VPN with the dynamic IP site. This functionality is facilitated by NHRP.

NOTE:
Always, the NHS address will be the private IP of the Tunnel Interface of the Hub.

R1(config)# interface Tunnel0 ←Get back into the Tunnel Interface
R1(config-if)# ip nhrp authentication NHRPkey ←Secret key for authenticating clients
R1(config-if)# ip nhrp map multicast dynamic ←Enables forwarding of multicast traffic across the tunnel.
R1(config-if)# ip nhrp network-id 100 ←Identifies the DMVPN cloud. All routers must have the same network ID.
R1(config-if)# tunnel key 100 ←OPTIONAL. Needed only on older IOS versions such as 12.3

Router-3 (SPOKE):

NHRP Configuration on Spoke sites has a few extra commands compared to the Hub.

R3(config)# interface Tunnel0 ←Get back into the Tunnel Interface
R3(config-if)# ip nhrp authentication NHRPkey ←Secret key for authenticating with Hub
R3(config-if)# ip nhrp map multicast dynamic ←Enables forwarding of multicast traffic across the tunnel.

R3(config-if)# **ip nhrp network-id 100**←Identifies the DMVPN cloud. All routers must have the same network ID.

R3(config-if)# **tunnel key 100**←<u>OPTIONAL</u>. Needed only on older IOS versions such as 12.3

R3(config-if)# **ip nhrp map 10.0.0.1 20.20.20.2** ←Map the NHS address (10.0.0.1 on Hub) with the Hub's WAN public IP (20.20.20.2)

R3(config-if)# **ip nhrp map multicast 20.20.20.2** ←Send multicast traffic to the Hub only. Hub will receive all multicast traffic (e.g routing protocol updates) and then send out updates to all the Spoke routers.

R3(config-if)# **ip nhrp nhs 10.0.0.1** ←Specify the NHS IP (this is always the private IP of the Tunnel Interface on Hub router)

R3(config-if)# **exit**

The above NHRP configuration will be exactly the same for all Spoke routers.

- **STEP 3: Configure Dynamic Routing**

Dynamic routing is a must in DMVPN topologies. In order to facilitate scalable deployments, configuring either EIGRP or OSPF will greatly enhance flexibility and configuration management. We will configure EIGRP in our example scenario. With EIGRP you must disable two default features of the protocol in order to allow proper routing advertisements between all devices in the topology.

First, you must disable "**Split Horizon**" on all routers. Split Horizon was designed to not allow a router to advertise a route out the same interface in which the route was originally learned. This conflicts in the case of the Tunnel interface because routers need to be able to make neighbors with the other routers on the same subnet (all Tunnel interfaces are in same subnet) and also re-advertise routes learned from one spoke to the next. The command is:

"*no ip split-horizon eigrp [AS_Number]*"

The second point to have in mind is "Next Hop Advertisement". Typically in EIGRP the next hop advertised is the router itself, but in DMVPN you want to make sure the spokes know about each other. In order to allow this to happen, you need to disable next-hop-self using

"no ip next-hop-self eigrp [AS_Number]"

Router-1 (HUB):

R1(config)# **router eigrp 90** ←Create an EIGRP instance with Autonomous System 90
R1(config-router)# **no auto-summary** ←Disable automatic summarization of subnets
R1(config-router)# **network 10.0.0.0 0.0.0.255** ←DMVPN Tunnel Interface Subnet
R1(config-router)# **network 192.168.1.0 0.0.0.255** ←Hub LAN subnet
R1(config-router)# **exit**

R1(config)# **interface Tunnel0** ←Get back into the Tunnel Interface
R1(config-if)# **no ip split-horizon eigrp 90** ←Disable Split Horizon
R1(config-if)# **no ip next-hop-self eigrp 90** ←Disable next hop self
R1(config-if)# **no ip redirects** ←Useful to have
R1(config-if)# **exit**

Router-3 (SPOKE):

R3(config)# **router eigrp 90** ←Create an EIGRP instance with Autonomous System 90
R3(config-router)# **no auto-summary** ←Disable automatic summarization of subnets
R3(config-router)# **network 10.0.0.0 0.0.0.255** ←DMVPN Tunnel Interface Subnet
R3(config-router)# **network 192.168.3.0 0.0.0.255** ←LAN subnet
R3(config-router)# **exit**
R3(config)# **interface Tunnel0** ←Get back into the Tunnel Interface
R3(config-if)# **no ip split-horizon eigrp 90** ←Disable Split Horizon
R3(config-if)# **no ip next-hop-self eigrp 90** ←Disable next hop self
R3(config-if)# **no ip redirects** ←Useful to have
R1(config-if)# **exit**

- **STEP 4: Configure IPSEC Protection**

I won't go into much detail in this Step since we have described both IPSEC Phase1 and Phase2 several times in previous examples. Let's see the configuration below:

Router-1 (HUB):

R1(config)# crypto isakmp policy 10
R1(config-isakmp)# encryption 3des
R1(config-isakmp)# hash md5
R1(config-isakmp)# authentication pre-share
R1(config-isakmp)# group 2
R1(config-isakmp)# exit

R1(config)# crypto isakmp key strongsecretkey address 0.0.0.0 0.0.0.0

R1(config)# crypto ipsec transform-set TRSET esp-3des esp-sha-hmac
R1(cfg-crypto-trans)# exit

R1(config)# crypto ipsec profile PROTECT-DMVPN
R1(ipsec-profile)# set transform-set TRSET
R1(ipsec-profile)# exit

R1(config)# interface Tunnel0
R1(config-if)# tunnel protection ipsec profile PROTECT-DMVPN

Router-3 (SPOKE):

R3(config)# crypto isakmp policy 10
R3(config-isakmp)# encryption 3des
R3(config-isakmp)# hash md5
R3(config-isakmp)# authentication pre-share
R3(config-isakmp)# group 2
R3(config-isakmp)# exit

R3(config)# crypto isakmp key strongsecretkey address 0.0.0.0 0.0.0.0 ←Spoke routers must allow also connections from any IP in order to form IPSEC VPN tunnels with other Spokes.

R3(config)# crypto ipsec transform-set TRSET esp-3des esp-sha-hmac
R3(cfg-crypto-trans)# exit

R3(config)# crypto ipsec profile PROTECT-DMVPN
R3(ipsec-profile)# set transform-set TRSET
R3(ipsec-profile)# exit

R3(config)# **interface Tunnel0**
R3(config-if)# **tunnel protection ipsec profile** PROTECT-DMVPN

This concludes the configuration of DMVPN. You can find a complete configuration example of the scenario above in Chapter 4, Section 4.1.10.

2.4 PPTP VPN

Point to Point Tunneling Protocol (PPTP) is a type of VPN that I have not mentioned in the previous Chapter. In fact, PPTP is a topic which I decided to include in the book at the last minute. At the beginning I believed that PPTP is kind of outdated and might not be useful for many people. However, I was wrong. I got several requests from people to include this kind of VPN connectivity because it is very useful in small to medium networks (SOHO etc), especially for providing an easy remote access solution to users with Microsoft OS computers. PPTP is natively supported by all current Windows operating systems without the need to install any additional software.

PPTP provides encrypted communication between a Client and a Server (client being a Microsoft computer and Server being a Cisco Router in our case). The PPTP encryption algorithms are not as strong as IPSEC or SSL VPN, but they offer a good security and privacy level. Moreover, PPTP uses PAP or CHAP protocols as the mechanisms to authenticate remote users connecting to the network.

Below we will describe how to configure a Cisco Router to work as a PPTP Server in order to terminate remote clients and provide them access to an internal network. PPTP is supported only on Routers. ASA firewalls cannot work as PPTP Servers.

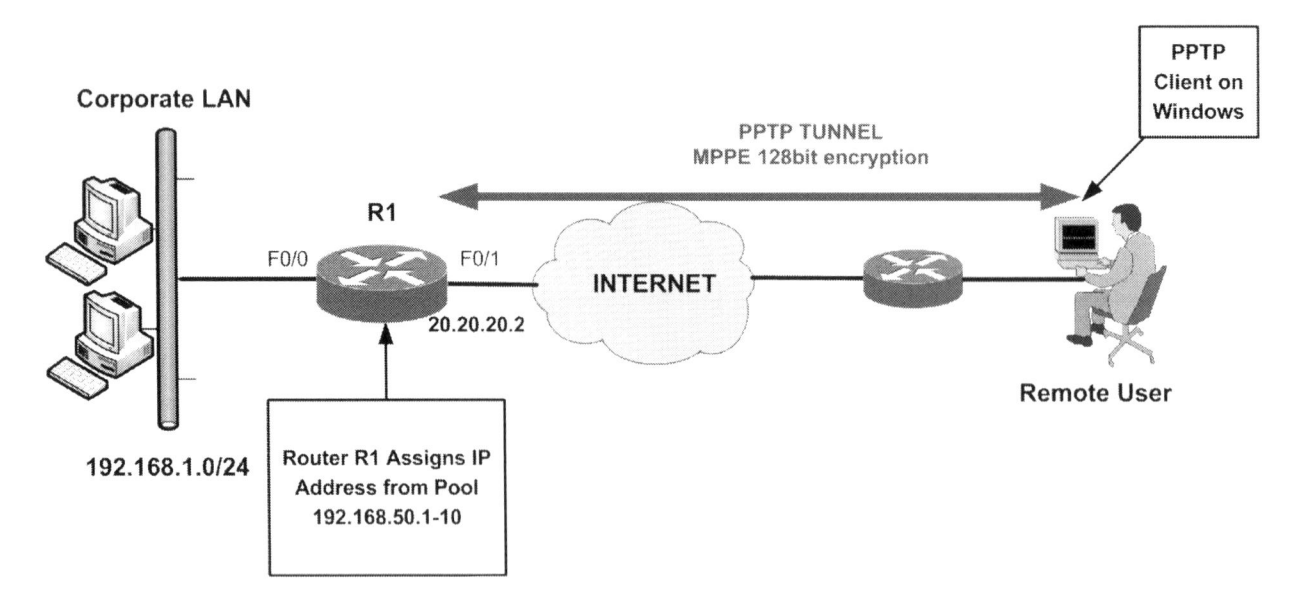

From diagram above, we have remote users who will be using PPTP to connect to their Corporate LAN network over the Internet. Router R1 will work as PPTP Server (also known as Virtual Private Dialup Network – VPDN Server) to accept connections from remote users and assign them an IP address from range 192.168.50.1-10. After that, remote users will have full access to the Corporate LAN. Let's see the configuration of R1 below:

<u>Router R1:</u>

- **STEP 1: Configure IP Pool to assign addresses to remote users**

Like all the other remote access VPN types, we need to have a pool of IP addresses to assign to remote users. In our case this pool will be in the range 192.168.50.1 to 192.168.50.10.

R1(config)#ip local pool pptp-pool 192.168.50.1 192.168.50.10

- **STEP 2: Configure the Virtual Private Dialup Network (VPDN)**

VPDN is the technology that was used originally in legacy dial-up networks. However, VPDN is used also for PPTP VPN connections.

!Enable and configure VPDN

R1(config)#vpdn enable← Enable the VPDN feature
R1(config)#vpdn-group 1← Create a VPDN group
R1(config-vpdn)#accept-dial ← Accept Dial In requests
R1(config-vpdn-acc-in)#protocol pptp ← Use the PPTP protocol
R1(config-vpdn-acc-in)#virtual-template 1← Attach virtual template 1 to this vpdn group
R1(config-vpdn-acc-in)#exit
R1(config-vpdn)#exit

- STEP 3: Configure a Virtual Interface for Terminating PPTP Tunnels

The following Virtual Interface will be used by all PPTP remote access users. This virtual interface template will assign an IP address to remote users from pool "**pptp-pool**" and will use the Microsoft Point to Point Encryption (mppe) with 128 bits. Also, authentication of remote users will be performed via "**ms-chap**" or "**ms-chap v2**". Notice also that this virtual interface must have an IP address in the same network range as the IP Pool of the remote users. When remote users access the router, virtual interfaces will be created which will be cloned from this virtual template.

R1(config)#interface Virtual-Template1
R1(config-if)#ip address 192.168.50.254 255.255.255.0← IP In the same range as the IP Pool
R1(config-if)#peer default ip address pool pptp-pool←Assign IP Pool "pptp-pool" to remote users
R1(config-if)#ppp encrypt mppe 128←Tunnel will be using the Microsoft Point to Point Encryption (mppe) with 128 bit key
R1(config-if)#ppp authentication ms-chap ms-chap-v2←Use either ms-chap or ms-chap2 for authentication of remote users.

- STEP 4: Create Username/Password Credentials for remote users

We have to create also credentials to be used by remote access users for authentication.

R1(config)#username remote1 password cisco123

- **STEP 5: Verification**

R1#show vpdn

```
PPTP Tunnel and Session Information Total tunnels 1 sessions 1
LocID Remote Name    State   Remote Address  Port  Sessions VPDN Group
2                   estabd  30.30.30.1      1659  1        1
LocID RemID TunID Intf   Username    State   Last Chg Uniq ID
2     6802  2     Vi3    remote1     estabd  00:02:05 1
```

As you can see above, there is one PPTP tunnel with remote IP **30.30.30.1** and username "**remote1**".

R1#show vpdn tunnel pptp all

```
PPTP Tunnel Information Total tunnels 1 sessions 1
Tunnel id 2, 1 active sessions
  Tunnel state is estabd, time since change 00:00:13
  Remote tunnel name is
    Internet Address 30.30.30.1, port 1659
  Local tunnel name is R1
    Internet Address 20.20.20.2, port 1723
  VPDN group: 1
  52 packets sent, 175 received, 2218 bytes sent, 20913 received
  Last clearing of "show vpdn" counters never
```

The above shows some more details about the established PPTP tunnel such as packets sent and received etc.

This concludes the configuration of PPTP VPN. You can find a complete configuration example of the scenario above in Chapter 4, Section 4.1.11.

Chapter 3   VPN Configuration on ASA Firewalls

This Chapter will focus on VPN configuration on Cisco ASA Firewall devices. The configurations here will be applicable for all ASA models in the 5500 and the new 5500-X series. For the configurations, we have used devices running ASA version 8.4(x) so some of the commands (e.g those about NAT) are different from older ASA versions (prior to 8.3). The VPN configurations below are also applicable for ASA versions 9.x and later.

3.1   Policy-Based VPN Configuration on Cisco ASA

3.1.1 Site-to-Site IPSEC VPN

This is the traditional IPSEC VPN (IKEv1 IPSEC) we have described in details in sections 1.2 and 2.1.1. Since we have already seen some theory and details about the IPSEC protocols and how they are used in VPN implementations, we will go directly to configuration steps for ASA firewalls.

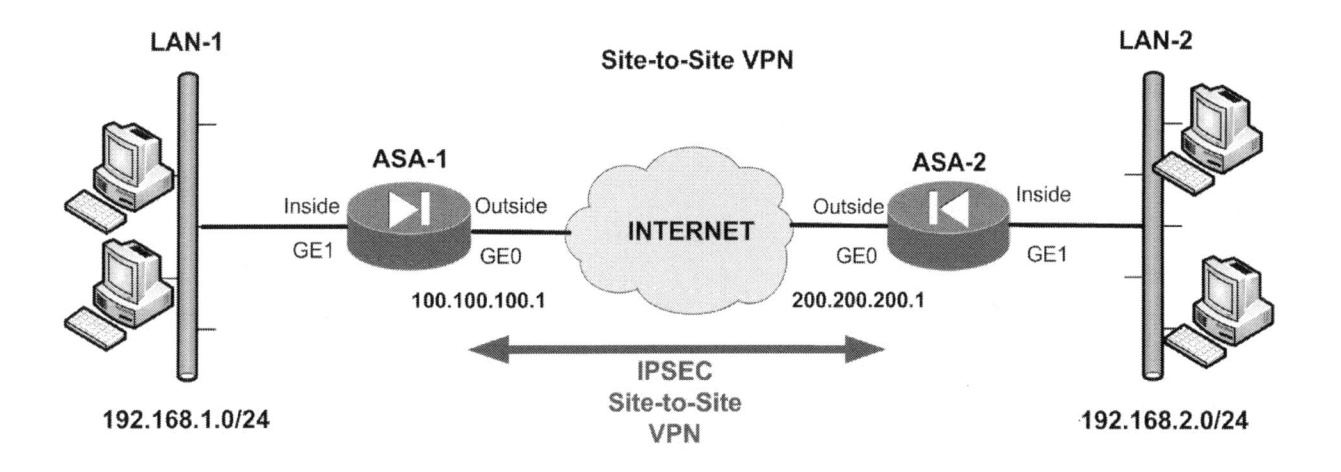

Our simple network topology above will help us to configure a site-to-site IPSEC VPN between two Cisco ASA devices. The configuration is the same for any ASA model.

- **STEP 1: Configure Interesting Traffic**

We need first to define the Interesting Traffic, that is, traffic that will be encrypted. Using Access-Lists (**Crypto ACL**) we can identify which traffic flow must be encrypted. In our example diagram above, we want all traffic flow between private networks 192.168.1.0/24 and 192.168.2.0/24 to be encrypted.

ASA 1:
ASA-1(config)# access-list VPN-ACL extended permit ip 192.168.1.0 255.255.255.0 192.168.2.0 255.255.255.0

ASA 2:
ASA-2(config)# access-list VPN-ACL extended permit ip 192.168.2.0 255.255.255.0 192.168.1.0 255.255.255.0

Notice that we have to configure the exact mirror access-list for each ASA firewall participating in the IPSEc VPN. The Crypto ACL needs to identify only outbound traffic. The **permit** statement in the ACL means that the specific traffic must be encrypted. Also, if you have additional internal networks behind the ASA that need to be transported in the VPN tunnel, then you must include them in the VPN-ACL as well.

NAT Exclusion

One important issue to consider is the case of using NAT on the firewall for normal Internet access. Because IPSEc does not work with NAT, we need to exclude the traffic to be encrypted from the NAT operation, as we did in IPSEC configuration on Routers in the previous Chapter. This means in our example that the Interesting Traffic in the Crypto ACL must not be translated.

The configuration below shows how to exclude VPN tunnel traffic from NAT operation. If you don't use NAT, then the configuration below is not needed.

ASA 1:
ASA-1(config)# object network obj-local
ASA-1(config-network-object)# subnet 192.168.1.0 255.255.255.0 ← LAN1
ASA-1(config-network-object)# exit

ASA-1(config)# object network obj-remote
ASA-1(config-network-object)# subnet 192.168.2.0 255.255.255.0 ← LAN2
ASA-1(config-network-object)# exit

ASA-1(config)# object network internal-lan ← This object will be used for PAT
ASA-1(config-network-object)# subnet 192.168.1.0 255.255.255.0
ASA-1(config-network-object)# exit

ASA-1(config)# nat (inside,outside) source static obj-local obj-local destination static obj-remote obj-remote ← Exclude traffic from LAN1 to LAN2 from NAT operation

ASA-1(config)# object network internal-lan
ASA-1(config-network-object)# nat (inside,outside) dynamic interface ← Configure Port Address Translation (PAT) using the outside ASA interface. This will perform dynamic NAT on internal LAN hosts so that they can access the Internet.

<u>ASA 2:</u>
ASA-2(config)# object network obj-local
ASA-2(config-network-object)# subnet 192.168.2.0 255.255.255.0 ← LAN2
ASA-2(config-network-object)# exit

ASA-2(config)# object network obj-remote
ASA-2(config-network-object)# subnet 192.168.1.0 255.255.255.0 ← LAN1
ASA-2(config-network-object)# exit

ASA-2(config)# object network internal-lan ← This object will be used for PAT
ASA-2(config-network-object)# subnet 192.168.2.0 255.255.255.0
ASA-2(config-network-object)# exit

ASA-2(config)# nat (inside,outside) source static obj-local obj-local destination static obj-remote obj-remote ← Exclude traffic from LAN2 to LAN1 from NAT operation

ASA-2(config)# object network internal-lan
ASA-2(config-network-object)# nat (inside,outside) dynamic interface ← Configure Port Address Translation (PAT) using the outside ASA interface. This will perform dynamic NAT on internal LAN hosts so that they can access the Internet.

- **STEP 2: Configure Phase 1 (ISAKMP - ikev1)**

The command format of the isakmp (known also as **ikev1**) policy in ASA Firewalls is the following:

ASA(config)# crypto ikev1 policy "*priority number*" ←Lower number means higher priority
ASA(config-ikev1-policy)# encryption {*aes |aes-192|aes-256|3des|des*}
ASA(config-ikev1-policy)# hash {*sha | md5*}
ASA(config-ikev1-policy)# authentication {*pre-share | rsa-sig*}
ASA(config-ikev1-policy)# group {*1 | 2 | 5 | 7*} ←DH Group
ASA(config-ikev1-policy)# lifetime "*seconds*" ←Up to 86400 seconds
ASA(config)# crypto ikev1 enable "*interface-name*" ←Attach the policy on an interface
ASA(config)# crypto isakmp identity address ←Identify the ASA with its address and not FQDN

Several isakmp policies can be configured to match different requirements from different IPSEc peers. The priority number uniquely identifies each policy. The lower the priority number, the higher the priority will be given to the specific policy.

The following example parameters can be used to create a strong ikev1 policy:
- Encryption **aes**
- Hash **sha**
- Authentication **pre-share**
- Group **2 or 5**
- Lifetime **86400** (the Security Association – SA will expire and renegotiate every 86400 sec)

The next thing we need to specify is the pre-shared key and the type of the VPN (site-to-site or Remote Access). These are configured by the **tunnel-group** command.

ASA(config)# tunnel-group *"peer IP address"* **type** {*ipsec-l2l | remote-access*}
ASA(config)# tunnel-group *"peer IP address"* **ipsec-attributes**
ASA(config-tunnel-ipsec)# ikev1 pre-shared-key *"key"*

Let's see the complete configuration on both firewalls for IPSEC Phase1 parameters:

ASA 1:
ASA-1(config)# crypto ikev1 policy 10
ASA-1(config-ikev1-policy)# authentication pre-share ← Use pre-shared key for auth
ASA-1(config-ikev1-policy)# encryption aes ← Use AES 128 bit encryption
ASA-1(config-ikev1-policy)# hash sha ← Use SHA for hashing
ASA-1(config-ikev1-policy)# group 2 ← Diffie-Hellman Group 2
ASA-1(config-ikev1-policy)# lifetime 86400 ← Lifetime of SA is 86400 seconds
ASA-1(config-ikev1-policy)# exit

ASA-1(config)# crypto ikev1 enable outside ← Enable the policy on "outside" interface
ASA-1(config)# crypto isakmp identity address

ASA-1(config)# tunnel-group 200.200.200.1 type ipsec-l2l ← Configure a tunnel with peer IP 200.200.200.1 which will be of type Lan-to-Lan
ASA-1(config)# tunnel-group 200.200.200.1 ipsec-attributes
ASA-1(config-tunnel-ipsec)# ikev1 pre-shared-key *somestrongkey* ← pre-shared key

ASA 2:
ASA-2(config)# crypto ikev1 policy 10
ASA-2(config-ikev1-policy)# authentication pre-share ← Use pre-shared key for auth
ASA-2(config-ikev1-policy)# encryption aes ← Use AES 128 bit encryption
ASA-2(config-ikev1-policy)# hash sha ← Use SHA for hashing
ASA-2(config-ikev1-policy)# group 2 ← Diffie-Hellman Group 2
ASA-2(config-ikev1-policy)# lifetime 86400 ← Lifetime of SA is 86400 seconds
ASA-2(config-ikev1-policy)# exit

ASA-2(config)# crypto ikev1 enable outside ← Enable the policy on "outside" interface
ASA-2(config)# crypto isakmp identity address

ASA-2(config)# tunnel-group 100.100.100.1 type ipsec-l2l ← Configure a tunnel with peer IP 100.100.100.1 which will be of type Lan-to-Lan
ASA-2(config)# tunnel-group 100.100.100.1 ipsec-attributes
ASA-2(config-tunnel-ipsec)# ikev1 pre-shared-key *somestrongkey* ← pre-shared key

- **STEP 3: Configure Phase 2 (IPSEc)**

As we did in IPSEC configuration on routers, for Phase2 we need to configure a "**Transform Set**" and "**Crypto Map**".

The command format of configuring a transform set is the following:

ASA(config)# crypto ipsec ikev1 transform-set *"name" "transform1" "transform2"*

The following transforms (protocols/algorithms) can be used in place of *transform1* and *transform2*:

Transform	Description
esp-des	ESP transform using DES cipher (56 bits)
esp-3des	ESP transform using 3DES cipher (168 bits)
esp-aes	ESP transform using AES-128 cipher
esp-aes-192	ESP transform using AES-192 cipher
esp-aes-256	ESP transform using AES-256 cipher
esp-md5-hmac	ESP transform using HMAC-MD5 authentication
esp-sha-hmac	ESP transform using HMAC-SHA authentication
esp-none	ESP with no authentication
esp-null	ESP with null encryption

After configuring a transform set on both IPSEc peers, we need to configure a **crypto map** which combines all Phase 2 IPSEc parameters. This crypto map is then attached to the firewall interface (usually "**outside**") on which the IPSEc will be established.

The command format of a crypto map is:

ASA(config)# crypto map "*name*" "*seq-num*" match address "*Crypto-ACL*" ←Assign the Crypto ACL which specifies the Interesting Traffic to be encrypted.
ASA(config)# crypto map "*name*" "*seq-num*" set peer "*Peer_IP_address*" ←Specify the remote peer IP address
ASA(config)# crypto map "*name*" "*seq-num*" set ikev1 transform-set "*Transform_set_name*" ←This is the transform set name configured above

ASA(config)# crypto map "*name*" interface "*interface-name*" ←Attach the map to an interface

The *seq-num* parameter in the crypto map is used to specify multiple map entries (with the same name) for cases where we have more than one IPSEc peers terminated on the same firewall. For example, if the above firewall is a Hub firewall in a Hub-and-Spoke VPN topology with 2 spokes, then there will be two crypto map entries with same "name" but different "sequence numbers".

Let's see the complete example configuration for both firewalls for Phase 2 setup:

ASA 1:

ASA-1(config)# crypto ipsec ikev1 transform-set TRSET esp-aes esp-md5-hmac
ASA-1(config)# crypto map VPNMAP 10 match address VPN-ACL
ASA-1(config)# crypto map VPNMAP 10 set peer 200.200.200.1
ASA-1(config)# crypto map VPNMAP 10 set ikev1 transform-set TRSET
ASA-1(config)# crypto map VPNMAP interface outside

ASA 2:

ASA-2(config)# crypto ipsec ikev1 transform-set TRSET esp-aes esp-md5-hmac
ASA-2(config)# crypto map VPNMAP 10 match address VPN-ACL
ASA-2(config)# crypto map VPNMAP 10 set peer 100.100.100.1
ASA-2(config)# crypto map VPNMAP 10 set ikev1 transform-set TRSET
ASA-2(config)# crypto map VPNMAP interface outside

- **STEP 4: Verify Encrypted Data Transfer**

With the three steps above we concluded the configuration of a site-to-site IPSEc VPN. An essential step though is to verify that everything is working fine and that our data is actually getting encrypted by the firewalls. There are two important commands that will help you verify if the tunnel is established and if data is bi-directionally encrypted between the IPSEc peers.

Verify that tunnel is established

The **show crypto isakmp sa** command verifies that the Security Association (SA) is established which means that the tunnel is up and running. Let's see an example output of this command below:

ASA-1# show crypto isakmp sa

```
IKEv1 SAs:

  Active SA: 1
   Rekey SA: 0 (A tunnel will report 1 Active and 1 Rekey SA during rekey)
Total IKE SA: 1

1   IKE Peer: 200.200.200.1
    Type  : L2L        Role  : initiator
    Rekey : no         State : MM_ACTIVE

There are no IKEv2 SAs
```

The important point to observe here is the **State : MM_ACTIVE**. This verifies that the IPSEc tunnel is established successfully.

Verify that data is bi-directionally encrypted

The **show crypto ipsec sa** command verifies that data is being encrypted and decrypted successfully by the firewall appliance, as shown below:

ASA-1# **show crypto ipsec sa**

```
interface: outside
  Crypto map tag: VPNMAP, seq num: 10, local addr: 100.100.100.1

    access-list VPN-ACL extended permit ip 192.168.1.0 255.255.255.0 192.168.2.0 255.255.255.0
    local ident (addr/mask/prot/port): (192.168.1.0/255.255.255.0/0/0)
    remote ident (addr/mask/prot/port): (192.168.2.0/255.255.255.0/0/0)
    current_peer: 200.200.200.1

    #pkts encaps: 8, #pkts encrypt: 8, #pkts digest: 8
    #pkts decaps: 8, #pkts decrypt: 8, #pkts verify: 8
    #pkts compressed: 0, #pkts decompressed: 0
    #pkts not compressed: 8, #pkts comp failed: 0, #pkts decomp failed: 0
  ---Output Omitted---
```

As shown above, we have packets being encrypted and decrypted (**pkts encrypt, pkts decrypt**) which shows that the IPSEC VPN tunnel is working as expected.

This concludes the configuration of a simple site-to-site IPSEC VPN using ASA firewalls. You can find a complete configuration of the scenario above in Chapter 4, Section 4.2.1.

3.1.1.1 Restricting IPSEC VPN Traffic between the Two Sites

By default, a site-to-site IPSEC VPN provides full network connectivity between the two LANs. This means that hosts in LAN1 can access all hosts in LAN2 and vice-versa. However, this might not be desirable is some situations. There are cases where we want hosts from one site to access only specific hosts of the other site and not the whole network.

In this section I will show you how to restrict IPSEC VPN traffic so that LAN-2 can access only two hosts on LAN-1 and not the whole network.

The key here is to disable the default command "**sysopt connection permit-vpn**". This command is enabled by default on Cisco ASA and its purpose is to exempt all IPSEC VPN traffic from Access List check on the outside ASA interface. This means that when the above command is enabled, all IPSEC VPN traffic is allowed to pass between the two sites without restricting anything. If we disable the command above, then we must explicitly allow the IPSEC traffic from the peer site on the outside

Access Control List of the ASA. Hence, we can apply fine-grained control of the IPSEC traffic between the two sites.

Note that IPSEC uses three protocols: **ESP**, **AH** and **IKE port UDP 500** (isakmp). Therefore we must allow those protocols on the outside Access List to reach the firewall interface. After that, we need also to explicitly allow which private hosts on LAN-1 can be accessed from LAN-2.

Let's see how to restrict IPSEC VPN traffic so that LAN-2 can access only two hosts (192.168.1.10 and 192.168.1.2) on LAN-1.

ASA-1
!First disable the IPSEC traffic exemption from Access List checks. This means that we must explicitly specify which VPN traffic is allowed to pass.
ASA-1(config)#no sysopt connection permit-vpn

!Now let's explicitly allow IPSEC traffic from LAN-2 to LAN-1. We need first to allow the three IPSEC Protocols from ASA-2 to ASA-1
ASA-1(config)#access-list outside_in extended permit esp host 200.200.200.1 host 100.100.100.1
ASA-1(config)#access-list outside_in extended permit ah host 200.200.200.1 host 100.100.100.1
ASA-1(config)#access-list outside_in extended permit udp host 200.200.200.1 host 100.100.100.1 eq isakmp

!Now allow access from LAN-2 to two hosts on LAN-1 only
ASA-1(config)#access-list outside_in extended permit ip 192.168.2.0 255.255.255.0 host 192.168.1.10
ASA-1(config)#access-list outside_in extended permit ip 192.168.2.0 255.255.255.0 host 192.168.1.2

!Apply the ACL to outside interface.
ASA-1(config)#access-group outside_in in interface outside

3.1.2 Hub-and-Spoke IPSEC VPN with Dynamic IP Spoke

In order to make things more interesting, we will discuss a Hub-and-Spoke scenario where we have one Spoke branch with static IP address and a second Spoke branch with dynamic IP address.

As we've described in hub-and-spoke VPN networks using Routers, basically a Hub-and-Spoke VPN network consists of several site-to-site IPSEC VPN tunnels between the Hub and each Spoke site. The configuration for the Spoke remote sites is the same as we've described in "Site-to-Site IPSEC VPN" section above so we won't talk about it again. However, the configuration of the Hub site firewall has a few differences as we will see below:

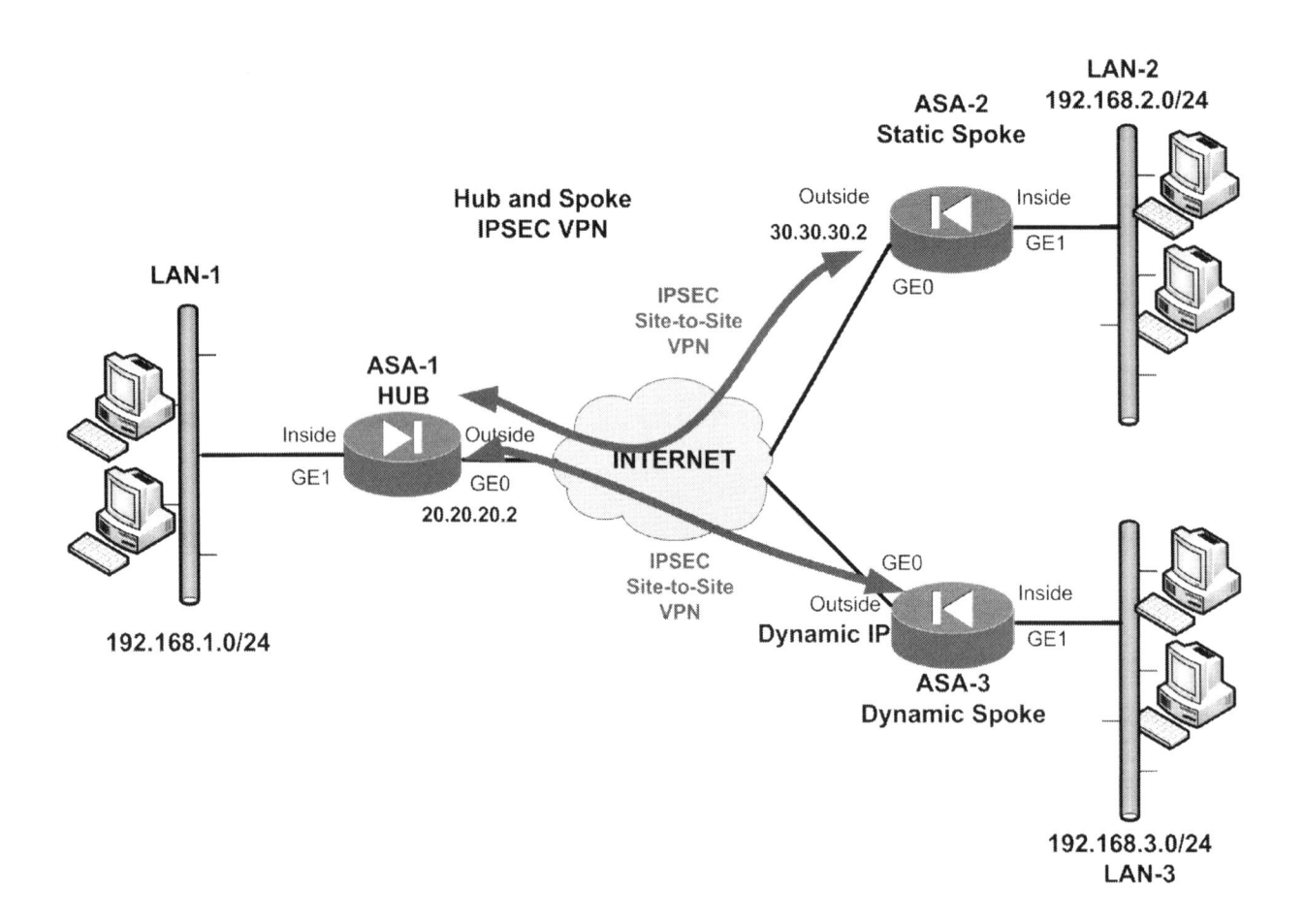

ASA-1 (HUB):

- **STEP 1: Configure Interesting Traffic and NAT Exemption**

!First identify the Interesting traffic to be encrypted. We need to have two crypto ACLs, one for each Spoke site.

ASA-1(config)# access-list VPN-ACL1 extended permit ip 192.168.1.0 255.255.255.0 192.168.2.0 255.255.255.0
ASA-1(config)# access-list VPN-ACL2 extended permit ip 192.168.1.0 255.255.255.0 192.168.3.0 255.255.255.0

!Then exclude the VPN Interesting traffic from the NAT operation

ASA-1(config)# object network obj-local
ASA-1(config-network-object)# subnet 192.168.1.0 255.255.255.0 ← Local LAN
ASA-1(config-network-object)# exit

ASA-1(config)# object network obj-remote1
ASA-1(config-network-object)# subnet 192.168.2.0 255.255.255.0 ← Spoke LAN2
ASA-1(config-network-object)# exit
ASA-1(config)# object network obj-remote2
ASA-1(config-network-object)# subnet 192.168.3.0 255.255.255.0 ← Spoke LAN3
ASA-1(config-network-object)# exit

ASA-1(config)# object network internal-lan ← This object will be used for PAT
ASA-1(config-network-object)# subnet 192.168.1.0 255.255.255.0
ASA-1(config-network-object)# exit

ASA-1(config)# nat (inside,outside) 1 source static obj-local obj-local destination static obj-remote1 obj-remote1 ← Exclude traffic from LAN1 to LAN2 from NAT operation

ASA-1(config)# nat (inside,outside) 2 source static obj-local obj-local destination static obj-remote2 obj-remote2 ← Exclude traffic from LAN1 to LAN3 from NAT operation

ASA-1(config)# object network internal-lan
ASA-1(config-network-object)# nat (inside,outside) dynamic interface ← Configure Port Address Translation (PAT) using the outside ASA interface. This will perform dynamic NAT on internal LAN hosts so that they can access the Internet.

- **STEP 2: Configure Phase 1 (ISAKMP - ikev1)**

!Configure Phase1 isakmp parameters

ASA-1(config)# crypto ikev1 policy 10
ASA-1(config-ikev1-policy)# authentication pre-share
ASA-1(config-ikev1-policy)# encryption 3des
ASA-1(config-ikev1-policy)# hash sha
ASA-1(config-ikev1-policy)# group 2
ASA-1(config-ikev1-policy)# lifetime 86400
ASA-1(config-ikev1-policy)# exit
ASA-1(config)# crypto ikev1 enable outside
ASA-1(config)# crypto isakmp identity address

!Configure static tunnel-group with the Static Spoke ASA-2

ASA-1(config)# tunnel-group 30.30.30.2 type ipsec-l2l ← Configure a static tunnel with ASA-2 (Static Spoke)
ASA-1(config)# tunnel-group 30.30.30.2 ipsec-attributes
ASA-1(config-tunnel-ipsec)# ikev1 pre-shared-key *secretkey1* ← pre-shared key with static spoke ASA-2

!Configure dynamic tunnel-group with the Dynamic Spoke ASA-3

ASA-1(config)# tunnel-group DefaultL2LGroup ipsec-attributes ← This is a special tunnel group with name "DefaultL2LGroup" which is used for dynamic IP spoke sites.

ASA-1(config-tunnel-ipsec)# ikev1 pre-shared-key *secretkey2* ← pre-shared key with dynamic spoke ASA-3

The default tunnel-group "**DefaultL2LGroup** " is used to match all branch sites having dynamic public IP address.

- **STEP 3: Configure Phase 2 (IPSEc)**

!Now Configure Phase2 Transform Set and Crypto Map. We need to create one dynamic crypto map and attach it to a static crypto map.

ASA-1(config)# crypto ipsec ikev1 transform-set TRSET esp-3des esp-md5-hmac

!Create a dynamic crypto map "DYNMAP" for the Dynamic IP Spoke (ASA3)
ASA-1(config)# crypto dynamic-map DYNMAP 10 match address VPN-ACL2
ASA-1(config)# crypto dynamic-map DYNMAP 10 set ikev1 transform-set TRSET

!Create a static crypto map "VPNMAP"
ASA-1(config)# crypto map VPNMAP 5 match address VPN-ACL1
ASA-1(config)# crypto map VPNMAP 5 set peer 30.30.30.2 ← Static IP Spoke
ASA-1(config)# crypto map VPNMAP 5 set ikev1 transform-set TRSET

!Attach the dynamic map to the static map
ASA-1(config)# crypto map VPNMAP 10 ipsec-isakmp dynamic DYNMAP

!Attach the static map to the outside interface
ASA-1(config)# crypto map VPNMAP interface outside

You can find a complete configuration of the scenario above in Chapter 4, Section 4.2.2.

3.1.2.1   *Spoke to Spoke Communication via the Hub ASA*

The Hub-and-Spoke network we have described above does not support communication between the two Spoke sites. Spokes can access only the Hub site. However, using "**VPN Hairpinning**" we can do a trick to allow remote Spoke sites to communicate between them through the central Hub site.

With VPN Hairpinning the Hub ASA firewall (ASA-1) will be configured to allow VPN traffic from spoke sites to enter and exit its "**outside**" interface so that spokes can communicate between them via the Hub. For example, private IP traffic from LAN-2 can travel through the VPN Tunnel between ASA-2 and ASA-1 and then exit from the same interface and get into the VPN Tunnel between ASA-1 and ASA-3 and finally reach LAN-3. This is called "**VPN Hairpinning**". The alternative way would be to configure a direct site-to-site IPSEC VPN between the two spokes (which is not very scalable if you have a lot of spokes).

In order to implement the functionality above you need to make changes to all ASA devices in the topology. Fortunately, you need to modify only the VPN Access Lists (for the Interesting Traffic) and also the NAT exemption rules. Moreover, the central Hub ASA requires also a command to allow traffic to enter and exit the same interface (**same-security-traffic permit intra-interface**). The rest of the configuration regarding the IPSEC VPN is not affected.

Let's see the configuration changes required on the three ASA devices in the topology above.

ASA-1 (HUB):

!Allow VPN traffic to enter and exit the same interface. This is essential in order for Spoke traffic to enter and exit the "outside" ASA interface to reach the other Spoke.

ASA-1(config)#same-security-traffic permit intra-interface

!Now modify the VPN ACL for LAN-1 to LAN-2 traffic to allow also LAN-3 to LAN-2 traffic
ASA-1(config)# access-list VPN-ACL1 extended permit ip 192.168.1.0 255.255.255.0 192.168.2.0 255.255.255.0
ASA-1(config)# access-list VPN-ACL1 extended permit ip 192.168.3.0 255.255.255.0 192.168.2.0 255.255.255.0 ← Allow also LAN-3 to LAN-2 traffic

!Now modify the VPN ACL for LAN-1 to LAN-3 traffic to allow also LAN-2 to LAN-3 traffic
ASA-1(config)# access-list VPN-ACL2 extended permit ip 192.168.1.0 255.255.255.0 192.168.3.0 255.255.255.0
ASA-1(config)# access-list VPN-ACL2 extended permit ip 192.168.2.0 255.255.255.0 192.168.3.0 255.255.255.0 ← Allow also LAN-2 to LAN-3 traffic

!The following are the Spoke LAN subnets
ASA-1(config)# object network obj-remote1
ASA-1(config-network-object)# subnet 192.168.2.0 255.255.255.0 ← Spoke LAN2
ASA-1(config-network-object)# exit

ASA-1(config)# object network obj-remote2
ASA-1(config-network-object)# subnet 192.168.3.0 255.255.255.0 ← Spoke LAN3
ASA-1(config-network-object)# exit

!Now Configure proper NAT Exemption
ASA-1(config)# nat (outside,outside) source static obj-remote1 obj-remote1 destination static obj-remote2 obj-remote2 ← Exclude traffic from LAN2 to LAN3 from NAT operation

ASA-1(config)# nat (outside,outside) source static obj-remote2 obj-remote2 destination static obj-remote1 obj-remote1 ← Exclude traffic from LAN3 to LAN2 from NAT operation

NOTE1: As you can see above, the NAT rule is using the same interface name "outside". This is because traffic between spokes is entering and exiting the same interface (outside) of the Hub ASA.

NOTE2: The other NAT Exemptions for traffic between LAN1 to LAN2/LAN3 are not shown here.

ASA-2 (Spoke1):

ASA-2(config)# object network obj-local
ASA-2(config-network-object)# subnet 192.168.2.0 255.255.255.0
ASA-2(config-network-object)# exit

ASA-2(config)# object network obj-hub
ASA-2(config-network-object)# subnet 192.168.1.0 255.255.255.0
ASA-2(config-network-object)# exit

ASA-2(config)# object network obj-spoke2
ASA-2(config-network-object)# subnet 192.168.3.0 255.255.255.0
ASA-2(config-network-object)# exit

!Modify the VPN ACL for LAN-2 to LAN-1 traffic to allow also LAN-2 to LAN-3 traffic
ASA-2(config)# access-list VPN-ACL extended permit ip 192.168.2.0 255.255.255.0 192.168.1.0 255.255.255.0
ASA-2(config)# access-list VPN-ACL extended permit ip 192.168.2.0 255.255.255.0 192.168.3.0 255.255.255.0 ← Allow also LAN-2 to LAN-3 traffic

!Now Configure proper NAT Exemption
ASA-2(config)# nat (inside,outside) source static obj-local obj-local destination static obj-hub obj-hub ← Exclude traffic from LAN2 to LAN1 from NAT operation

ASA-2(config)# nat (inside,outside) source static obj-local obj-local destination static obj-spoke2 obj-spoke2 ← Exclude traffic from LAN2 to LAN3 from NAT operation

ASA-3 (Spoke2):

ASA-3(config)# object network obj-local
ASA-3(config-network-object)# subnet 192.168.3.0 255.255.255.0
ASA-3(config-network-object)# exit

ASA-3(config)# object network obj-hub
ASA-3(config-network-object)# subnet 192.168.1.0 255.255.255.0
ASA-3(config-network-object)# exit

ASA-3(config)# object network obj-spoke1
ASA-3(config-network-object)# subnet 192.168.2.0 255.255.255.0
ASA-3(config-network-object)# exit

!Modify the VPN ACL for LAN-3 to LAN-1 traffic to allow also LAN-3 to LAN-2 traffic
ASA-3(config)# access-list VPN-ACL extended permit ip 192.168.3.0 255.255.255.0 192.168.1.0 255.255.255.0

ASA-3(config)# access-list VPN-ACL extended permit ip 192.168.3.0 255.255.255.0 192.168.2.0 255.255.255.0 ← Allow also LAN-3 to LAN-2 traffic

!Now Configure proper NAT Exemption
ASA-3(config)# **nat (inside,outside) source static obj-local obj-local destination static obj-hub obj-hub** ← Exclude traffic from LAN3 to LAN1 from NAT operation

ASA-3(config)# **nat (inside,outside) source static obj-local obj-local destination static obj-spoke1 obj-spoke1** ← Exclude traffic from LAN3 to LAN2 from NAT operation

3.1.3 IPSEC VPN between Cisco ASA and Cisco Router

As we have said at the beginning of this book, one of the advantages of IPSEC protocol is that it can be used to create VPNs between different types of devices and even between different vendors. This is because IPSEC is an IETF standard protocol and is supported almost on all VPN capable devices.

Here we will see a configuration of site-to-site VPN between a Cisco Router and ASA firewall. The important point to remember is that the configuration on the Router or ASA is the same as site-to-site IPSEC we have described in previous sections where we had only Routers or ASA devices in the network. So let's see quickly the configuration commands based on the network diagram below:

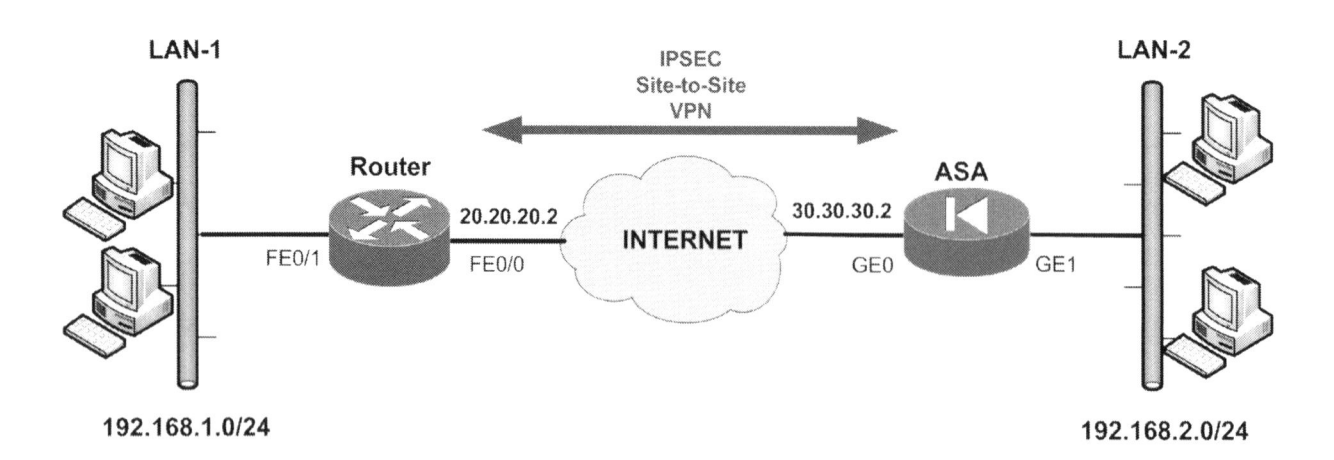

ROUTER:

! Configure Interesting Traffic

R1(config)# ip access-list extended VPN-ACL

R1(config-ext-nacl)# permit ip 192.168.1.0 0.0.0.255 192.168.2.0 0.0.0.255

! Configure NAT exclusion

R1(config)# ip access-list extended NAT-ACL

R1(config-ext-nacl)# deny ip 192.168.1.0 0.0.0.255 192.168.2.0 0.0.0.255

R1(config-ext-nacl)# permit ip 192.168.1.0 0.0.0.255 any

! Enable the NAT functionality on FE0/1 (inside) and FE0/0 (outside) interfaces

R1(config)# ip nat inside source list NAT-ACL interface FastEthernet0/0 overload

R1(config)# interface FastEthernet0/0

R1(config-if)# ip nat outside

R1(config)# interface FastEthernet0/1

R1(config-if)# ip nat inside

! Configure IPSEC Phase1.

R1(config)# crypto isakmp policy 10

R1(config-isakmp)# encryption 3des

R1(config-isakmp)# hash md5

R1(config-isakmp)# authentication pre-share

R1(config-isakmp)# group 2

R1(config-isakmp)# exit

! In this scenario we will use the new features, "keyring" and "isakmp profile"

R1(config)# crypto keyring ASAVPNKEY

R1(conf-keyring)# pre-shared-key address 30.30.30.2 key secretkey1

R1(config)# crypto isakmp profile staticL2L

R1(conf-isa-prof)# keyring ASAVPNKEY

R1(conf-isa-prof)# match identity address 30.30.30.2 255.255.255.255

R1(conf-isa-prof)# exit

! Configure IPSEC Phase2.

R1(config)# crypto ipsec transform-set TRSET esp-3des esp-md5-hmac

R1(cfg-crypto-trans)# exit

R1(config)# crypto map VPNMAP 10 ipsec-isakmp

R1(config-crypto-map)# set peer 30.30.30.2

R1(config-crypto-map)# set transform-set TRSET

R1(config-crypto-map)# set isakmp-profile staticL2L

R1(config-crypto-map)# match address VPN-ACL

R1(config-crypto-map)# exit

!Attach the crypto map above to the WAN outside interface (FE0/0) of the router

R1(config)# interface FastEthernet0/0

R1(config-if)# crypto map VPNMAP

ASA:

! Interesting traffic and NAT

ASA-1(config)# access-list VPN-ACL extended permit ip 192.168.2.0 255.255.255.0 192.168.1.0 255.255.255.0

ASA-1(config)# object network obj-local
ASA-1(config-network-object)# subnet 192.168.2.0 255.255.255.0
ASA-1(config-network-object)# exit

ASA-1(config)# object network obj-remote
ASA-1(config-network-object)# subnet 192.168.1.0 255.255.255.0
ASA-1(config-network-object)# exit

ASA-1(config)# object network internal-lan
ASA-1(config-network-object)# subnet 192.168.2.0 255.255.255.0
ASA-1(config-network-object)# exit

```
ASA-1(config)# nat (inside,outside) source static obj-local obj-local destination static obj-remote obj-remote

ASA-1(config)# object network internal-lan
ASA-1(config-network-object)# nat (inside,outside) dynamic interface
```

! IPSEC Phase1

```
ASA-1(config)# crypto ikev1 policy 10
ASA-1(config-ikev1-policy)# authentication pre-share
ASA-1(config-ikev1-policy)# encryption 3des
ASA-1(config-ikev1-policy)# hash md5
ASA-1(config-ikev1-policy)# group 2
ASA-1(config-ikev1-policy)# lifetime 86400
ASA-1(config-ikev1-policy)# exit

ASA-1(config)# crypto ikev1 enable outside
ASA-1(config)# crypto isakmp identity address

ASA-1(config)# tunnel-group 20.20.20.2 type ipsec-l2l
ASA-1(config)# tunnel-group 20.20.20.2 ipsec-attributes
ASA-1(config-tunnel-ipsec)# ikev1 pre-shared-key secretkey1
```

! IPSEC Phase2

```
ASA-1(config)# crypto ipsec ikev1 transform-set TRSET esp-3des esp-md5-hmac
ASA-1(config)# crypto map VPNMAP 10 match address VPN-ACL
ASA-1(config)# crypto map VPNMAP 10 set peer 20.20.20.2
ASA-1(config)# crypto map VPNMAP 10 set ikev1 transform-set TRSET
ASA-1(config)# crypto map VPNMAP interface outside
```

You can find a complete configuration example of the scenario above in Chapter 4, Section 4.2.3.

3.1.4 Remote Access IPSEC VPN

This is the same scenario as 2.1.3 but now the remote users get access to their corporate LAN via an ASA firewall device. Let's move quickly to the configuration details based on the network below.

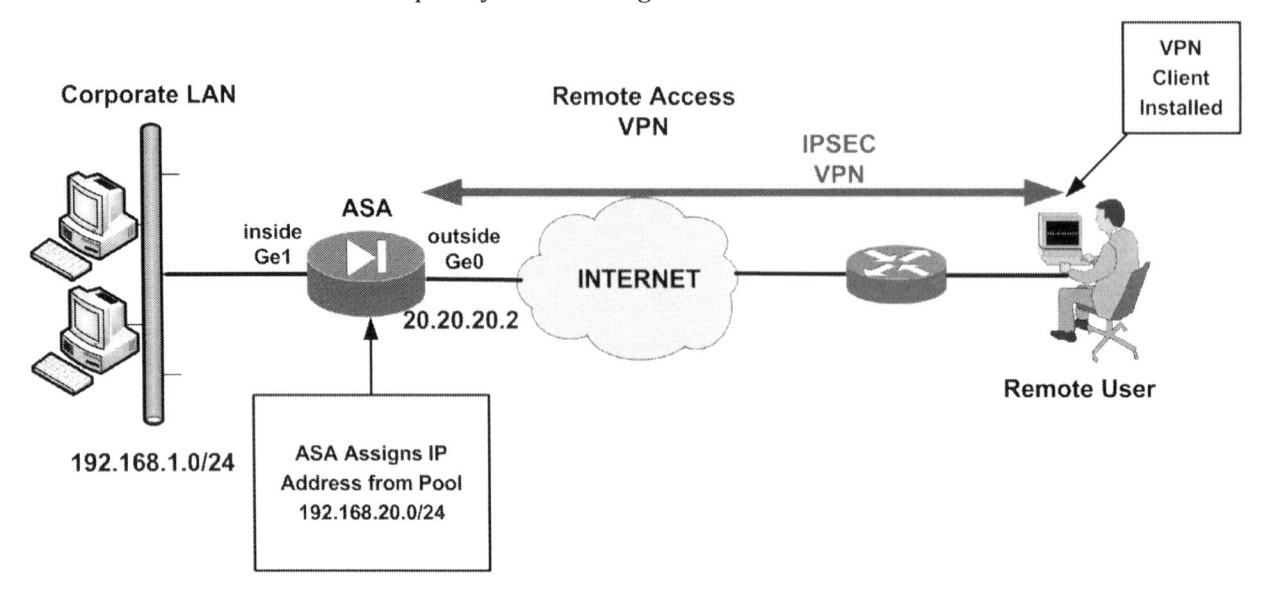

<u>ASA:</u>
- **STEP 1: Configure VPN Pool**

!First configure an IP address pool that will be used to assign IP addresses to remote users

ASA(config)# ip local pool vpnpool 192.168.20.1-192.168.20.254

- **STEP 2: Configure Split-Tunneling (Optional)**

Once the Remote Access VPN is established, the remote user by default will not be able to access anything else on the Internet, except the Corporate LAN network. This behavior can be altered by configuring the "**split tunneling**" feature on the Firewall, which however is not recommended for security purposes. However, if you want to allow users to access the Internet and also access the Corporate LAN network, you must configure a Split-Tunnel Access Control List.

ASA(config)# access-list splittunnel standard permit 192.168.1.0 255.255.255.0

Traffic from the remote users towards the network specified in the split-tunnel ACL (192.168.1.0/24) will pass through the VPN tunnel. All other traffic from the remote user will go to the Internet.

- **STEP 3: Configure NAT Exemption**

Similarly with site-to-site VPN, we need to exclude from NAT the traffic flow from our Internal LAN network (192.168.1.0/24) towards the Remote Users (192.168.20.0/24).

ASA(config)# object network obj-local
ASA(config-network-object)# subnet 192.168.1.0 255.255.255.0
ASA(config-network-object)# exit

ASA(config)# object network obj-vpnpool
ASA(config-network-object)# subnet 192.168.20.0 255.255.255.0
ASA(config-network-object)# exit

ASA(config)# nat (inside,outside) source static obj-local obj-local destination static obj-vpnpool obj-vpnpool

- **STEP 4: Configure Group Policy**

The Group Policy allows you to separate different remote access users into groups with different attributes. For example System Administrators can be assigned in a group having 24-hours VPN access, while normal remote user can be in a different group with 9am-5pm VPN access. The Group Policy also provides DNS or WINS server addresses, connection filtering, idle timeout settings etc.

ASA(config)# group-policy vpn-clients-policy internal
ASA(config)# group-policy vpn-clients-policy attributes
ASA(config-group-policy)# vpn-idle-timeout 30
ASA(config-group-policy)# dns-server value 192.168.1.5
ASA(config-group-policy)# split-tunnel-policy tunnelspecified
ASA(config-group-policy)# split-tunnel-network-list value splittunnel

Assume that all remote users will use the same group policy, with the name "**vpn-clients-policy**" as configured above. This policy assigns an internal DNS server address so that users can resolve internal domains. It sets also the idle timeout to 30 minutes. Also, under the Group Policy we assign the Split-Tunnel ACL (splittunnel) which will dictate which traffic will pass through the tunnel from the remote clients.

NOTE:

Under Group Policy you can configure also a **VPN Filter** with a purpose of restricting access from remote users to certain IPs or Ports in the Corporate LAN. For example, assume that you want remote VPN users to access only a specific Internal Server at port 80 and disallow anything else. You must configure a filter ACL and apply it under the Group Policy:

ASA(config)# access-list VPN-FILTER-ACL extended permit tcp 192.168.20.0 255.255.255.0 host 192.168.1.10 eq 80 ← Allow access from remote users to server 192.168.1.10 port 80

ASA(config)# group-policy vpn-clients-policy attributes
ASA(config-group-policy)# vpn-filter value VPN-FILTER-ACL

- **STEP 5: Configure Usernames for Authentication**

When remote users connect to the ASA using the VPN client, they will be presented with a login screen in order to authenticate with the firewall. We need therefore to create username/password credentials for authentication. In our scenario we will use locally configured users for this.

ASA(config)# username vpnuser password test123

- **STEP 6: Configure IPSEC Phase 1 (ikev1)**

This is the same as site-to-site VPN.

ASA(config)# crypto ikev1 policy 10
ASA(config-ikev1-policy)# encryption 3des
ASA(config-ikev1-policy)# hash sha
ASA(config-ikev1-policy)# authentication pre-share
ASA(config-ikev1-policy)# group 2
ASA(config-ikev1-policy)# lifetime 86400
ASA(config-ikev1-policy)# exit
ASA(config)# crypto ikev1 enable outside
ASA(config)# crypto isakmp identity address

- **STEP 7: Configure IPSEC Phase 2**

This Step also has similarities with site-to-site VPNs. We need an IPSEC transform set which will specify the encryption and authentication protocols for the Remote Access VPN. Also, we need to configure a dynamic crypto map which will be assigned to a static crypto map.

ASA(config)# **crypto ipsec ikev1 transform-set TRSET esp-3des esp-md5-hmac**
ASA(config)# **crypto dynamic-map** outside_dyn_map **10 set ikev1 transform-set TRSET**

ASA(config)# **crypto map IPSEC 10 ipsec-isakmp dynamic** outside_dyn_map
ASA(config)# **crypto map IPSEC interface outside**

- **STEP 8: Configure a Tunnel Group for Remote Access**

The tunnel group configuration is the heart of remote access VPN. It binds together the Group Policy configured before, the IP pool assignment, the pre-shared key etc.

ASA(config)# **tunnel-group** remotevpn **type remote-access**
ASA(config)# **tunnel-group** remotevpn **general-attributes**
ASA(config-tunnel-general)# **address-pool vpnpool** ← Assign the IP pool from Step 1
ASA(config-tunnel-general)# **default-group-policy vpn-clients-policy** ← Assign Group Policy with name "vpn-clients-policy" from Step 4.
ASA(config-tunnel-general)#**exit**

ASA(config)# **tunnel-group** remotevpn **ipsec-attributes**
ASA(config-tunnel-ipsec)# **ikev1 pre-shared-key** cisco123

The Tunnel Group name "**remotevpn**" and the Pre-Shared Key (Password) "**cisco123**"are important here because we will have to specify them in the "**Group Authentication**" settings when configuring the VPN client software as shown in picture below:

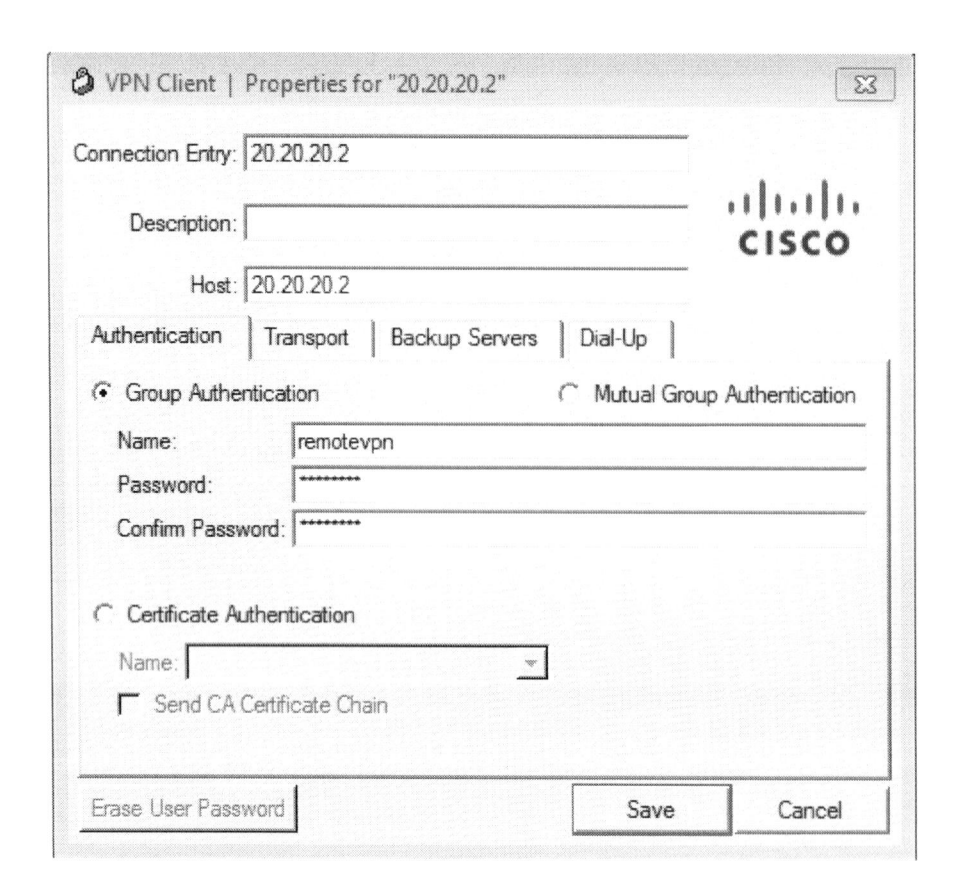

- **STEP 9: Verification**

Now let's verify that everything is working fine. Assuming the remote user authenticates successfully, we will see the following on ASA:

ASA# show crypto ipsec sa

interface: outside

  Crypto map tag: outside_dyn_map, seq num: 10, local addr: 20.20.20.2

  local ident (addr/mask/prot/port): (0.0.0.0/0.0.0.0/0/0)

  remote ident (addr/mask/prot/port): (192.168.20.1/255.255.255.255/0/0)

  current_peer: 195.12.101.240, username: ***vpnuser***

  dynamic allocated peer ip: ***192.168.20.1***

  #pkts encaps: 4, #pkts encrypt: 4, #pkts digest: 4

  #pkts decaps: 4, #pkts decrypt: 4, #pkts verify: 4

…Output Omitted…

As shown above, the remote user "**vpnuser**" has received a dynamic allocated IP of **192.168.20.1**. Also, we have packets encrypted and decrypted.

You can find a complete configuration of the scenario above in Chapter 4, Section 4.2.4.

3.1.5 Hub-and-Spoke and Remote Access VPN on same device

In section 2.1.4 we have seen an interesting scenario where we had a mixture of static site-to-site VPN, dynamic site-to-site VPN and remote access IPSEC VPN all terminated on the same Central Hub Router. We will examine the same scenario here using Cisco ASA firewalls instead of routers.

As depicted in the network diagram below, we have a central Hub ASA device and two remote branch sites (Spokes). One of the remote sites (ASA2) has a static Public IP whereas the second remote site (ASA3) has a dynamic Public IP. Moreover, we want also to accommodate remote users who want to connect securely with remote access VPN client to the central network (LAN-1).

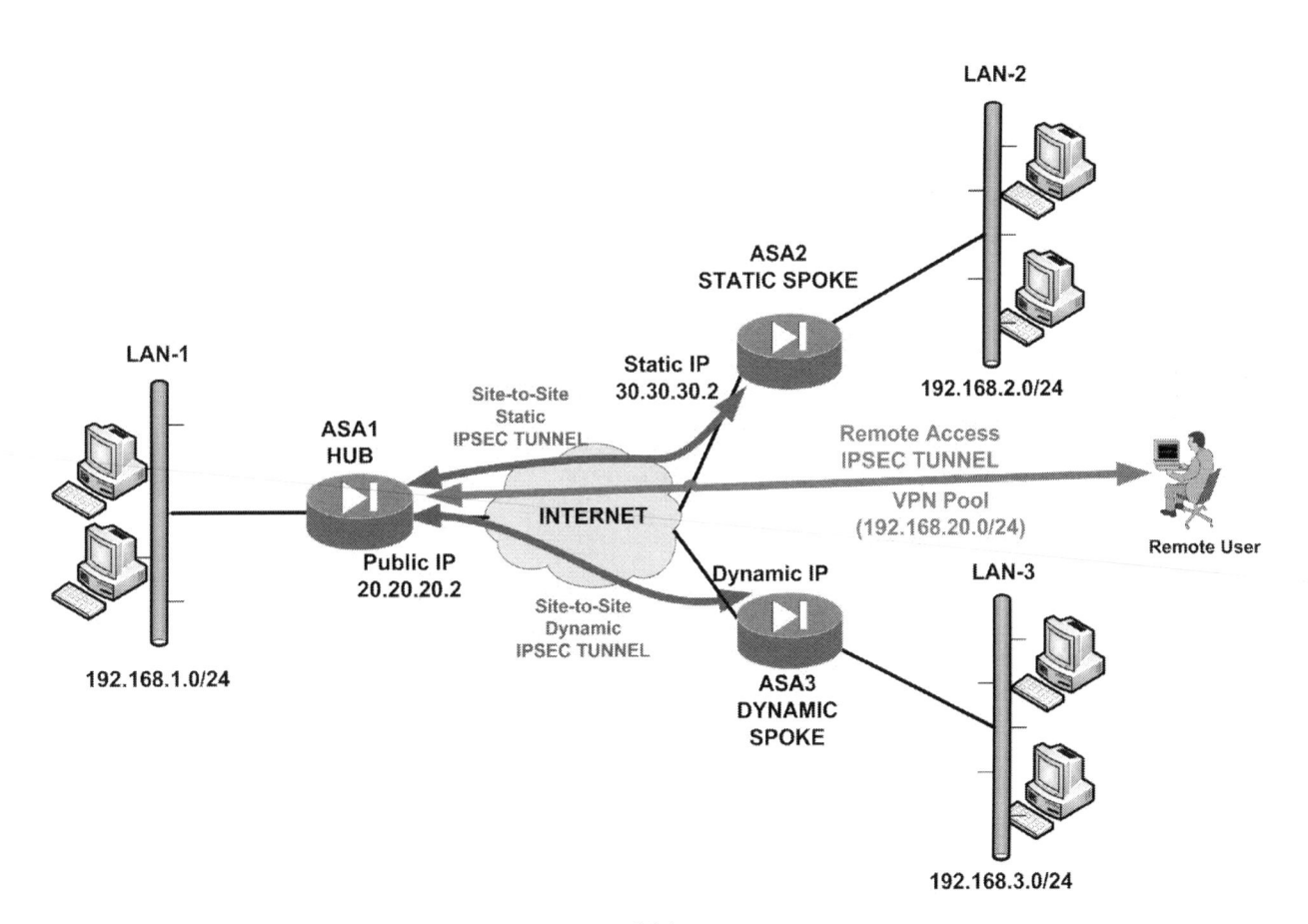

We will see how to configure the HUB device (ASA1) only since the configuration for the remote spoke devices is the same as 3.1.1.

ASA1 - HUB:

- **STEP 1: Configure Interesting Traffic and NAT Exemption**

!First identify the Interesting traffic to be encrypted. We need to have two crypto ACLs, one for each Spoke site.

ASA-1(config)# access-list VPN-ACL1 extended permit ip 192.168.1.0 255.255.255.0 192.168.2.0 255.255.255.0
ASA-1(config)# access-list VPN-ACL2 extended permit ip 192.168.1.0 255.255.255.0 192.168.3.0 255.255.255.0

!Then exclude the VPN Interesting traffic from the NAT operation

ASA-1(config)# object network obj-local
ASA-1(config-network-object)# subnet 192.168.1.0 255.255.255.0 ← Local LAN
ASA-1(config-network-object)# exit

ASA-1(config)# object network obj-remote1
ASA-1(config-network-object)# subnet 192.168.2.0 255.255.255.0 ← Spoke LAN2
ASA-1(config-network-object)# exit

ASA-1(config)# object network obj-remote2
ASA-1(config-network-object)# subnet 192.168.3.0 255.255.255.0 ← Spoke LAN3
ASA-1(config-network-object)# exit

ASA-1(config)# object network obj-vpnpool ← IP Pool for remote access clients
ASA-1(config-network-object)# subnet 192.168.20.0 255.255.255.0
ASA-1(config-network-object)# exit

ASA-1(config)# object network internal-lan ← This object will be used for PAT
ASA-1(config-network-object)# subnet 192.168.1.0 255.255.255.0
ASA-1(config-network-object)# exit

ASA-1(config)# nat (inside,outside) 1 source static obj-local obj-local destination static obj-remote1 obj-remote1 ← Exclude traffic from LAN1 to LAN2 from NAT operation

ASA-1(config)# nat (inside,outside) 2 source static obj-local obj-local destination static obj-remote2 obj-remote2 ← Exclude traffic from LAN1 to LAN3 from NAT operation

ASA-1(config)# nat (inside,outside) 3 source static obj-local obj-local destination static obj-vpnpool obj-vpnpool no-proxy-arp route-lookup ← Exclude traffic from LAN1 towards the remote access VPN clients pool

ASA-1(config)# object network internal-lan
ASA-1(config-network-object)# nat (inside,outside) dynamic interface ← Configure Port Address Translation (PAT) using the outside ASA interface. This will perform dynamic NAT on internal LAN hosts so that they can access the Internet.

- **STEP 2: Configure VPN Pool and Split Tunneling if needed**

!First configure an IP address pool that will be used to assign IP addresses to remote users

ASA-1(config)# ip local pool vpnpool 192.168.20.1-192.168.20.254
!Configure split tunneling if needed
ASA-1(config)# access-list splittunnel standard permit 192.168.1.0 255.255.255.0

- **STEP 3: Configure Phase 1 (ISAKMP - ikev1)**

!Configure Phase1 isakmp parameters

ASA-1(config)# crypto ikev1 policy 10
ASA-1(config-ikev1-policy)# authentication pre-share
ASA-1(config-ikev1-policy)# encryption 3des
ASA-1(config-ikev1-policy)# hash sha
ASA-1(config-ikev1-policy)# group 2
ASA-1(config-ikev1-policy)# lifetime 86400
ASA-1(config-ikev1-policy)# exit
ASA-1(config)# crypto ikev1 enable outside
ASA-1(config)# crypto isakmp identity address

!Configure static tunnel-group with the Static Spoke ASA-2

ASA-1(config)# tunnel-group 30.30.30.2 type ipsec-l2l ← Configure a static tunnel with ASA-2 (Static Spoke)
ASA-1(config)# tunnel-group 30.30.30.2 ipsec-attributes
ASA-1(config-tunnel-ipsec)# ikev1 pre-shared-key *secretkey1* ← pre-shared key with static spoke ASA-2

!Configure dynamic tunnel-group with the Dynamic Spoke ASA-3

ASA-1(config)# tunnel-group DefaultL2LGroup ipsec-attributes ← This is a special tunnel group with name "DefaultL2LGroup" which is used for dynamic IP spoke sites.

ASA-1(config-tunnel-ipsec)# ikev1 pre-shared-key *secretkey2* ← pre-shared key with dynamic spoke ASA-3

The default tunnel-group "DefaultL2LGroup " is used to match all branch sites having dynamic public IP address.

!Configure a tunnel-group for the remote access vpn clients

ASA-1(config)# tunnel-group remotevpn type remote-access
ASA-1(config)# tunnel-group remotevpn general-attributes
ASA-1(config-tunnel-general)# address-pool vpnpool
ASA-1(config-tunnel-general)# default-group-policy vpn-clients-policy ← Assign Group Policy "vpn-clients-policy" (see 3.1.4 Step 4)
ASA-1(config-tunnel-general)#exit

ASA-1(config)# tunnel-group remotevpn ipsec-attributes
ASA-1(config-tunnel-ipsec)# ikev1 pre-shared-key cisco123

- STEP 4: Configure Phase 2 (IPSEc)

!Now Configure Phase2 Transform Set and Crypto Map. We need to create one dynamic crypto map and attach it to a static crypto map.

ASA-1(config)# crypto ipsec ikev1 transform-set TRSET esp-3des esp-md5-hmac

!Create a dynamic crypto map "DYNMAP" with two entries. One entry for the dynamic Spoke Site and another entry for the vpn clients.

ASA-1(config)# crypto dynamic-map DYNMAP 5 match address VPN-ACL2
ASA-1(config)# crypto dynamic-map DYNMAP 5 set ikev1 transform-set TRSET
ASA-1(config)# crypto dynamic-map DYNMAP 10 set ikev1 transform-set TRSET

(Note: DYNMAP 5 is for the dynamic Spoke Site and DYNMAP 10 is for the VPN clients)

!Create a static crypto map "VPNMAP" with two entries. One entry for the static Spoke Site and another entry to attach the dynamic map above.

ASA-1(config)# crypto map VPNMAP 5 match address VPN-ACL1
ASA-1(config)# crypto map VPNMAP 5 set peer 30.30.30.2
ASA-1(config)# crypto map VPNMAP 5 set ikev1 transform-set TRSET

!This entry is to attach the dynamic map to the static map
ASA-1(config)# crypto map VPNMAP 10 ipsec-isakmp dynamic DYNMAP

!Attach the static map to the outside interface
ASA-1(config)# crypto map VPNMAP interface outside

- **STEP 5: Configure Users for Authentication of VPN clients**

!Configure username/password for the VPN clients.
ASA-1(config)# username vpnuser password test123

You can find a complete configuration of the scenario above in Chapter 4, Section 4.2.5.

3.1.5.1 Enable Remote Users to Access Spoke Sites through the Hub

The scenario in 3.1.5 above supports communication of remote access users with the central Hub network only. However, there are circumstances where we want to allow access from remote users to Spoke networks as well. We can do this via the Hub site. That is, remote users connect as normal to the Hub ASA and then they can access any Spoke site through the established Hub-and-Spoke VPN tunnels. This functionality uses the "VPN Hairpinning" concept as described in paragraph 3.1.2.1.

Next I will show you how to allow access from remote users to LAN2 in addition to the central LAN1 network (based on the diagram above). As explained in section 3.1.2.1 previously, we need to change VPN Access Lists and NAT Exemption rules. Let's see the configuration required on ASA1 and ASA2 for this:

ASA1 (HUB)

!Allow VPN traffic to enter and exit the same interface. This is essential in order for Remote Users' traffic to enter and exit the "outside" ASA interface to reach the Spoke.

ASA-1(config)#same-security-traffic permit intra-interface

!Now modify the VPN ACL for LAN-1 to LAN-2 traffic to allow also remote access IP pool traffic
ASA-1(config)# access-list VPN-ACL1 extended permit ip 192.168.1.0 255.255.255.0 192.168.2.0 255.255.255.0
ASA-1(config)# access-list VPN-ACL1 extended permit ip 192.168.20.0 255.255.255.0 192.168.2.0 255.255.255.0 ← Remote Access IP Pool to LAN-2

!Configure also split tunneling for LAN2 network
ASA-1(config)#access-list splittunnel standard permit 192.168.2.0 255.255.255.0

!The following are the Spoke and remote users subnets
ASA-1(config)# object network obj-remote1
ASA-1(config-network-object)# subnet 192.168.2.0 255.255.255.0 ← Spoke LAN2
ASA-1(config-network-object)# exit

ASA-1(config)# object network obj-vpnpool
ASA-1(config-network-object)# subnet 192.168.20.0 255.255.255.0 ← Remote Users pool
ASA-1(config-network-object)# exit

!Now Configure proper NAT Exemption
ASA-1(config)# nat (outside,outside) source static obj-vpnpool obj-vpnpool destination static obj-remote1 obj-remote1 ← Exclude traffic from remote pool to LAN2

NOTE1: As you can see above, the NAT rule is using the same interface name "**outside**". This is because traffic between remote access users and Spoke site is entering and exiting the same interface (outside) of the Hub ASA.

NOTE2: The other NAT exemptions rules are not shown above.

<u>ASA-2 (Spoke):</u>

ASA-2(config)# object network obj-local
ASA-2(config-network-object)# subnet 192.168.2.0 255.255.255.0
ASA-2(config-network-object)# exit

ASA-2(config)# object network obj-hub
ASA-2(config-network-object)# subnet 192.168.1.0 255.255.255.0
ASA-2(config-network-object)# exit

ASA-2(config)# object network obj-vpnpool
ASA-2(config-network-object)# subnet 192.168.20.0 255.255.255.0
ASA-2(config-network-object)# exit

!Modify the VPN ACL for LAN-2 to LAN-1 traffic to allow also LAN-2 to remote users pool
ASA-2(config)# access-list VPN-ACL extended permit ip 192.168.2.0 255.255.255.0 192.168.1.0 255.255.255.0
ASA-2(config)# access-list VPN-ACL extended permit ip 192.168.2.0 255.255.255.0 192.168.20.0 255.255.255.0 ← Allow LAN-2 to remote users IP pool

!Now Configure proper NAT Exemption
ASA-2(config)# nat (inside,outside) source static obj-local obj-local destination static obj-hub obj-hub ← Exclude traffic from LAN2 to LAN1 from NAT operation

ASA-2(config)# nat (inside,outside) source static obj-local obj-local destination static obj-vpnpool obj-vpnpool ← Exclude traffic from LAN2 to remote pool from NAT operation

3.1.6 Site-to-Site IPSEC VPN with failover using backup ISP

Moving away from the traditional configurations we have seen so far, let's now see an interesting and important scenario: Using a backup ISP for providing IPSEC VPN failover. Although the specific setup we have here refers to site-to-site topology, once you learn how to configure this it can be easily applied to Hub-and-Spoke topologies as well (remember that a Hub-and-Spoke topology is like having multiple site-to-site VPN topologies).

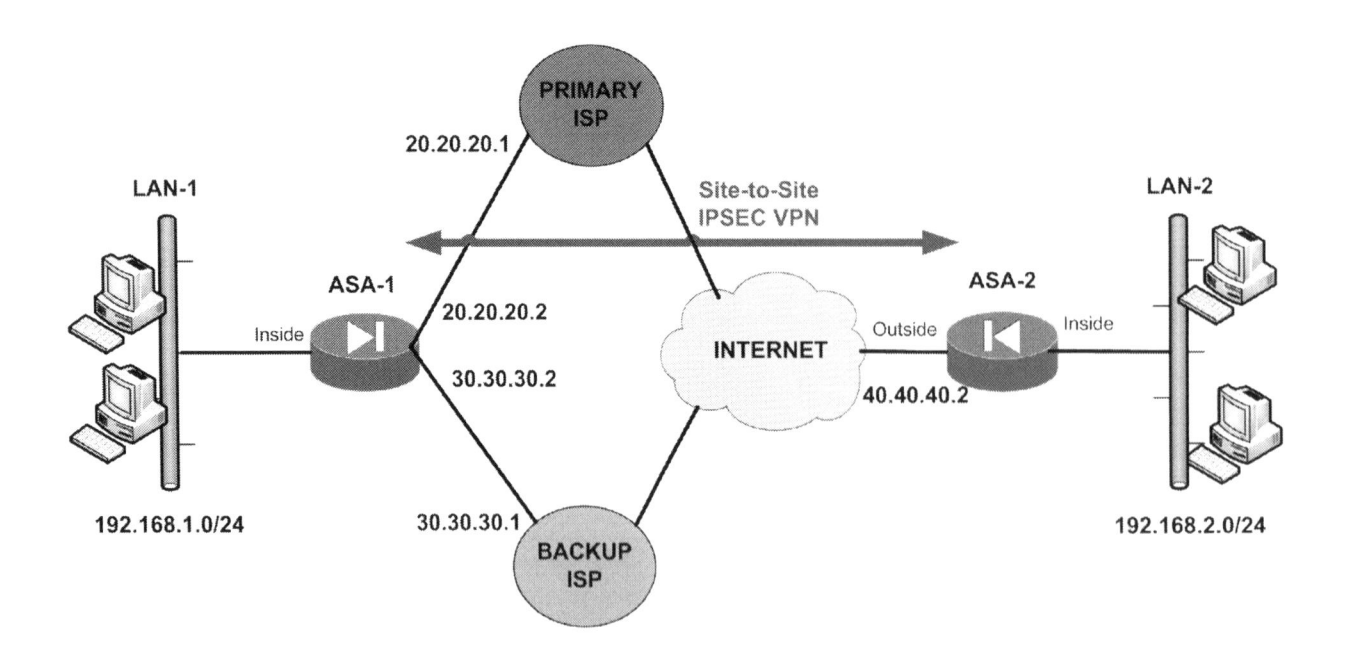

From the network diagram above, we have a central ASA device (ASA-1) connected to two ISPs. The Primary ISP connection has public IP 20.20.20.2 and the Backup ISP connection has public IP 30.30.30.2. If the primary connection fails for any reason, the ASA device will automatically switch over to the backup connection. We will see also how to configure this backup functionality on ASA1.

Moreover, we need to establish a site-to-site tunnel with a remote branch site (ASA2). In order to take advantage of the backup ISP capability of ASA1, we have to configure also a failover mechanism for the IPSEC VPN tunnel on ASA2. This is accomplished by configuring two "VPN peer IPs" on ASA2 as we will see below.

Since we have described in detail the site-to-site IPSEC VPN configuration before, we will not provide too many detailed comments here. Let's see the configuration on both ASA devices:

- **STEP 1: Configure the Two ISP Interfaces**

First let's configure the two external interfaces of Cisco ASA which will connect to the two ISPs.

ASA 1:

```
ASA-1(config)# interface GigabitEthernet0
ASA-1(config-if)# nameif primary-isp
ASA-1(config-if)# security-level 0
ASA-1(config-if)# ip address 20.20.20.2 255.255.255.0
ASA-1(config-if)# exit
```

```
ASA-1(config)# interface GigabitEthernet1
ASA-1(config-if)# nameif backup-isp
ASA-1(config-if)# security-level 0
ASA-1(config-if)# ip address 30.30.30.2 255.255.255.0
ASA-1(config-if)# exit
```

- **STEP 2: Configure Backup ISP Failover on ASA-1**

Now let's configure ISP Failover on ASA-1. This is achieved via "**Static Route Tracking**" and "**SLA monitor**" features.

Static Route Tracking

When you configure a static route on the security appliance, the route remains permanent in the routing table. The only way for the static route to get removed from the routing table is when the associated ASA interface goes physically down. In all other cases, such as for example when the remote default gateway goes down, the ASA will keep sending packets to its gateway router without knowing that it is actually down.

From ASA version 7.2 and later, the **Static Route Tracking** feature was introduced. The ASA tracks the availability of static routes by sending ICMP echo request packets through the primary static route path and waits for replies. If the primary path is down, a secondary path is used. This feature is useful when you want to implement Backup-ISP redundancy, as we will see below.

In our scenario above, two default static routes will be configured (one for each ISP) which will use the "**track**" feature. The primary ISP path will be tracked using ICMP echo requests. If an echo reply is not received within a predefined period, the backup static route will be used. Note however that the scenario above is suitable only for outbound communication (that is, from the inside network towards the Internet).

Configuring Static Route Tracking

1. Use the "**sla monitor**" command to specify the monitoring protocol (e.g ICMP), the target address to track (e.g ISP gateway router) and the tracking timers.
2. Use the "**sla monitor schedule**" command to schedule the monitoring process (usually the monitoring process is configured to run "**forever**" but duration and start times are configurable).
3. Define the primary static route to be tracked using the "**route**" command with the "**track**" option.
4. Define the backup static route and set its metric higher than the primary static route.

Let's see the configuration for backup ISP redundancy.

ASA 1:

ASA-1(config)# sla monitor 100 ← Define SLA_ID 100

! Use ICMP echo protocol for tracking the Primary ISP Gateway IP 20.20.20.1
ASA-1(config-sla-monitor)# type echo protocol ipIcmpEcho 20.20.20.1 interface primary-isp

ASA-1(config-sla-monitor-echo)# timeout 3000 ← Define timeout 3000 milliseconds (3 sec)
ASA-1(config-sla-monitor-echo)# frequency 5 ← track target 5 times
ASA-1(config-sla-monitor-echo)# exit

ASA-1(config)# sla monitor schedule 100 life forever start-time now ← Schedule the monitoring process SLA_ID 100 to start now and run forever

ASA-1(config)# track 10 rtr 100 reachability ← Associate a Track_ID 10 with the SLA_ID 100

ASA-1(config)# route primary-isp 0.0.0.0 0.0.0.0 20.20.20.1 1 track 10 ← Associate the Track_ID 10 to the primary ISP static route. Define also a metric 1 for this route.

ASA-1(config)# route backup-isp 0.0.0.0 0.0.0.0 30.30.30.1 254 ← Define the backup ISP static route with a higher route metric of 254

In the scenario above, the firewall appliance will be tracking the primary ISP gateway router (20.20.20.1). If an echo reply is not received within 3 sec (timeout 3000 milliseconds) and the process is repeated 5 times (frequency 5), the primary default route is considered down and therefore the secondary backup route will be used.

- **STEP 3: Configure IPSEC VPN on ASA-1**

Now let's get into the actual IPSEC VPN configuration.

<u>ASA 1:</u>

!Configure Interesting Traffic and NAT Exemption

ASA-1(config)# access-list VPN-ACL extended permit ip 192.168.1.0 255.255.255.0 192.168.2.0 255.255.255.0

ASA-1(config)# object network obj-local
ASA-1(config-network-object)# subnet 192.168.1.0 255.255.255.0 ← LAN1
ASA-1(config-network-object)# exit

ASA-1(config)# object network obj-remote
ASA-1(config-network-object)# subnet 192.168.2.0 255.255.255.0 ← LAN2
ASA-1(config-network-object)# exit

ASA-1(config)# nat (inside,primary-isp) source static obj-local obj-local destination static obj-remote obj-remote ← Exclude traffic from LAN1 to LAN2 from NAT operation when going through the Primary ISP link

ASA-1(config)# nat (inside,backup-isp) source static obj-local obj-local destination static obj-remote obj-remote ← Exclude traffic from LAN1 to LAN2 from NAT operation when going through the Backup ISP link

!Now configure the rest of IPSEC

ASA-1(config)# crypto ikev1 policy 10
ASA-1(config-ikev1-policy)# authentication pre-share
ASA-1(config-ikev1-policy)# encryption 3des
ASA-1(config-ikev1-policy)# hash sha
ASA-1(config-ikev1-policy)# group 2
ASA-1(config-ikev1-policy)# lifetime 86400
ASA-1(config-ikev1-policy)# exit

!Enable the isakmp policies on both Primary and Backup ISP interfaces

ASA-1(config)# crypto ikev1 enable primary-isp
ASA-1(config)# crypto ikev1 enable backup-isp

ASA-1(config)# crypto isakmp identity address

ASA-1(config)# tunnel-group 40.40.40.2 type ipsec-l2l ← Configure a tunnel with remote site peer IP 40.40.40.2
ASA-1(config)# tunnel-group 40.40.40.2 ipsec-attributes
ASA-1(config-tunnel-ipsec)# ikev1 pre-shared-key *somestrongkey*

ASA-1(config)# crypto ipsec ikev1 transform-set TRSET esp-3des esp-md5-hmac
ASA-1(config)# crypto map VPNMAP 10 match address VPN-ACL
ASA-1(config)# crypto map VPNMAP 10 set peer 40.40.40.2
ASA-1(config)# crypto map VPNMAP 10 set ikev1 transform-set TRSET

!Attach the crypto maps to both ISP interfaces
ASA-1(config)# crypto map VPNMAP interface primary-isp
ASA-1(config)# crypto map VPNMAP interface backup-isp

The above concludes the configuration of ASA-1. Let's now move on to configuring ASA-2

- **STEP 4: Configure IPSEC VPN on ASA-2**

We will see just a snapshot of the configuration of ASA-2 since the rest of the configuration is the same as described in site-to-site VPN scenario in section 3.1.1

<u>ASA 2:</u>

The only things that change on ASA-2 configuration are the **Crypto Map** and **Tunnel Group** commands.

On Crypto Map you have to specify two peer IP addresses for ASA-1. One will be the Primary link IP (20.20.20.2) of ASA-1 and the other one will be the Backup link IP (30.30.30.2) of ASA-1.

ASA-2(config)# crypto ipsec ikev1 transform-set TRSET esp-3des esp-md5-hmac
ASA-2(config)# crypto map VPNMAP 10 match address VPN-ACL
ASA-2(config)# crypto map VPNMAP 10 set peer 20.20.20.2 30.30.30.2 ←Specify two peer IP addresses for ASA-1. If the first IP is not reachable, tunnel will be formed with the second IP.
ASA-2(config)# crypto map VPNMAP 10 set ikev1 transform-set TRSET
ASA-2(config)# crypto map VPNMAP interface outside

Also, we have to specify two Tunnel Groups, one for each Public IP of ASA-1.

ASA-2(config)# tunnel-group 20.20.20.2 type ipsec-l2l ← Configure a tunnel with the Primary ISP Link of ASA-1
ASA-2(config)# tunnel-group 20.20.20.2 ipsec-attributes
ASA-2(config-tunnel-ipsec)# ikev1 pre-shared-key *somestrongkey*

ASA-2(config)# tunnel-group 30.30.30.2 type ipsec-l2l ← Configure a tunnel with the Backup ISP Link of ASA-1
ASA-2(config)# tunnel-group 30.30.30.2 ipsec-attributes
ASA-2(config-tunnel-ipsec)# ikev1 pre-shared-key *somestrongkey*

You can find a complete configuration of the scenario above in Chapter 4, Section 4.2.6.

3.1.7 Site-to-Site IPSEC VPN with Duplicate Subnets –Example1

The next two scenarios are very special. They are rare in real world and they are also a little bit difficult to configure. We are talking about a site-to-site IPSEC VPN where the two LAN networks at each site have the same subnet (Duplicate or Overlapping Networks).

The examples we have seen so far assume that LAN1 and LAN2 are different networks (usually we use 192.168.1.0/24 for LAN-1 and 192.168.2.0/24 for LAN-2). However, there are cases where we want to create a VPN tunnel between two LAN networks which use the same private subnet. Maybe your company merged with another company which happened to use the same private LAN subnet. Let's see how to configure such a topology using the diagram below as an example:

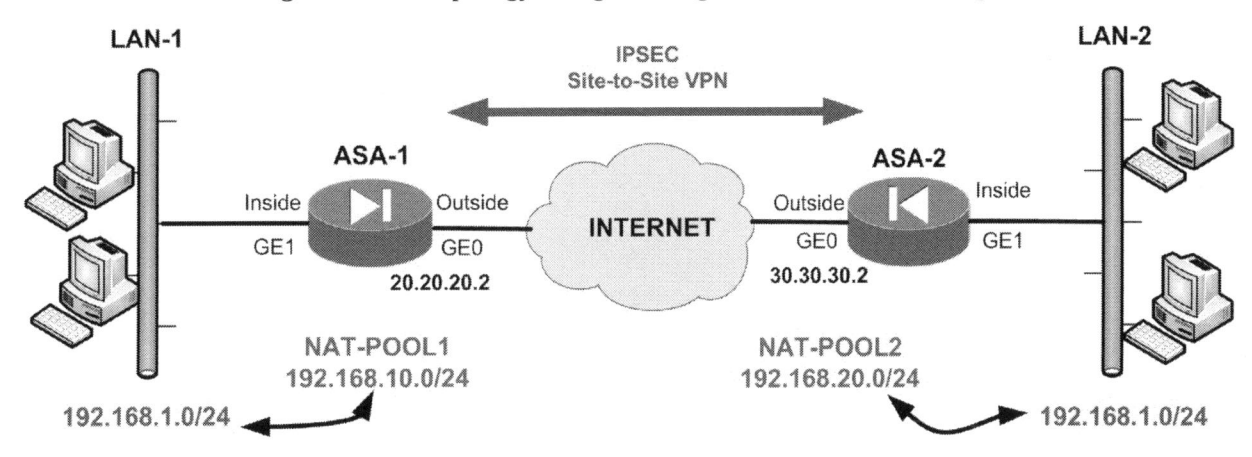

As shown from the network above, both LAN-1 and LAN-2 use subnet 192.168.1.0/24. If we create a VPN tunnel between the two sites, hosts in LAN-1 won't be able to communicate with hosts in LAN-2 (and vice-versa). In order to allow this communication, we need to configure policy NAT.

With policy NAT we will achieve the following:

- When hosts in LAN-1 want to access hosts in LAN-2 they will be translated to 192.168.10.0/24 (NAT-POOL1).
- When hosts in LAN-2 want to access hosts in LAN-1 they will be translated to 192.168.20.0/24 (NAT-POOL2).
- Hosts in LAN-1 will see hosts in LAN-2 as 192.168.20.0/24.
- Hosts in LAN-2 will see hosts in LAN-1 as 192.168.10.0/24.
- Therefore we will create two different "mapped" networks (NAT-POOL1, NAT-POOL2) in order to eliminate the duplicate subnets.

Let's see the configuration for both ASA devices (we will show only the configuration which is different from the traditional site-to-site VPN we have seen before).

ASA 1:

```
!Create the required network objects which will be used in NAT
ASA-1(config)# object network obj-local
ASA-1(config-network-object)# subnet 192.168.1.0 255.255.255.0   ← LAN1
ASA-1(config-network-object)# exit

ASA-1(config)# object network NAT-POOL1
ASA-1(config-network-object)# subnet 192.168.10.0 255.255.255.0
ASA-1(config-network-object)# exit

ASA-1(config)# object network NAT-POOL2
ASA-1(config-network-object)# subnet 192.168.20.0 255.255.255.0
ASA-1(config-network-object)# exit
```

!Configure Policy NAT. The local LAN (obj-local) will be translated to NAT-POOL1 when the destination network is NAT-POOL2

ASA-1(config)# nat (inside,outside) source static obj-local NAT-POOL1 destination static NAT-POOL2 NAT-POOL2

ASA-1(config)# object network obj-local
ASA-1(config-network-object)# nat (inside,outside) dynamic interface ← Configure Port Address Translation (PAT) using the outside ASA interface. This will perform dynamic NAT on internal LAN hosts so that they can access the Internet.

!Configure VPN Interesting Traffic which will be between the "mapped" networks (NAT-POOL1, NAT-POOL2). Remember to use the following VPN-ACL in a Crypto Map.
ASA-1(config)# access-list VPN-ACL extended permit ip 192.168.10.0 255.255.255.0 192.168.20.0 255.255.255.0

The rest of the configuration of ASA-1 is the same as traditional site-to-site IPSEC VPN.

<u>ASA 2:</u>

!Create the required network objects which will be used in NAT
ASA-2(config)# object network obj-local
ASA-2(config-network-object)# subnet 192.168.1.0 255.255.255.0 ← LAN2
ASA-2(config-network-object)# exit

ASA-2(config)# object network NAT-POOL1
ASA-2(config-network-object)# subnet 192.168.10.0 255.255.255.0
ASA-2(config-network-object)# exit

ASA-2(config)# object network NAT-POOL2
ASA-2(config-network-object)# subnet 192.168.20.0 255.255.255.0
ASA-2(config-network-object)# exit

!Configure Policy NAT. The local LAN (obj-local) will be translated to NAT-POOL2 when the destination network is NAT-POOL1

ASA-2(config)# nat (inside,outside) source static obj-local NAT-POOL2 destination static NAT-POOL1 NAT-POOL1

ASA-2(config)# object network obj-local
ASA-2(config-network-object)# nat (inside,outside) dynamic interface ← Configure Port Address Translation (PAT) using the outside ASA interface. This will perform dynamic NAT on internal LAN hosts so that they can access the Internet.

!Configure VPN Interesting Traffic which will be between the "mapped" networks (NAT-POOL2, NAT-POOL1). Remember to use the following VPN-ACL in a Crypto Map.
ASA-2(config)# access-list VPN-ACL extended permit ip 192.168.20.0 255.255.255.0 192.168.10.0 255.255.255.0

The rest of the configuration of ASA-2 is the same as traditional site-to-site IPSEC VPN.

NOTE:

Let's say you are a user in LAN-1 and you want to access host 192.168.1.35 located in LAN-2. To do this you must use 192.168.20.35 as destination IP instead of the actual IP which is 192.168.1.35.

Similarly, assume you are a user in LAN-2 and you want to access host 192.168.1.72 located in LAN-1. To do this you must access the host in LAN-1 as 192.168.10.72 instead of 192.168.1.72.

VERIFICATION

Let's see some output from the first ASA-1 device:

ASA1# show crypto ipsec sa

```
interface: outside
  Crypto map tag: VPNMAP, seq num: 10, local addr: 20.20.20.2

    access-list VPN-ACL extended permit ip 192.168.10.0 255.255.255.0 192.168.20.0 255.255.255.0
    local ident (addr/mask/prot/port): (192.168.10.0/255.255.255.0/0/0)
    remote ident (addr/mask/prot/port): (192.168.20.0/255.255.255.0/0/0)
    current_peer: 30.30.30.2

    #pkts encaps: 13, #pkts encrypt: 13, #pkts digest: 13
    #pkts decaps: 13, #pkts decrypt: 13, #pkts verify: 13
    #pkts compressed: 0, #pkts decompressed: 0
```

As shown from the output above, traffic between 192.168.10.0 and 192.168.20.0 is being encrypted. You can find a complete configuration of the scenario above in Chapter 4, Section 4.2.7.

3.1.8 Site-to-Site IPSEC VPN with Duplicate Subnets –Example2

This is our second example scenario with Duplicate Subnets in a site-to-site VPN. Here, assume that you are in control of one site of the network only. That is, you are the owner of ASA-1 but you don't have control on ASA-2 (see diagram below). This is common in situations where you have several customers or external partners that want to have access to your network via IPSEC VPN. In such cases we must do all the required configuration on our own ASA device in order to accommodate any duplicate subnet issues. See diagram below for more information:

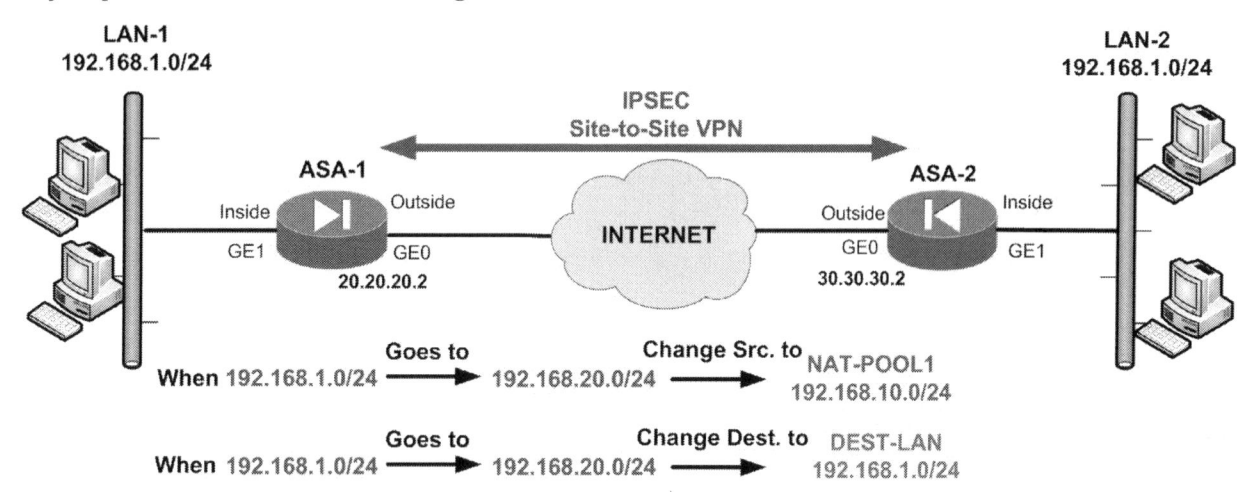

NOTE:

- Hosts in LAN-1 will access hosts in LAN-2 as 192.168.20.0/24 (NAT-POOL2).
- Hosts in LAN-2 will access hosts in LAN-1 as 192.168.10.0/24 (NAT-POOL1).

To implement the required functionality we must configure two special NAT configurations on ASA-1. The first NAT configuration will be a **Policy NAT** and the second NAT configuration will be a **Destination NAT**. The two NAT configurations will achieve the following:

- The **Policy NAT** will translate the Source IP of network 192.168.1.0/24 (LAN1) to 192.168.10.0/24 (NAT-POOL1) only when the Destination Network is 192.168.20.0/24 (NAT-POOL2).
- Outbound packets from ASA-1 going to 192.168.20.0/24 will have their Destination IP changed to 192.168.1.0/24. This is achieved with the **Destination NAT** statement.

Let's see the configuration for both ASA devices (we will show only the configuration which is different from the traditional site-to-site VPN we have seen before).

ASA 1:

!Create the required network objects which will be used in NAT
ASA-1(config)# **object network** LOCAL-LAN
ASA-1(config-network-object)# **subnet 192.168.1.0 255.255.255.0** ← LAN1
ASA-1(config-network-object)# **exit**
ASA-1(config)# **object network** DEST-LAN
ASA-1(config-network-object)# **subnet 192.168.1.0 255.255.255.0** ← LAN2
ASA-1(config-network-object)# **exit**

ASA-1(config)# **object network** NAT-POOL1
ASA-1(config-network-object)# **subnet 192.168.10.0 255.255.255.0**
ASA-1(config-network-object)# **exit**

ASA-1(config)# **object network** NAT-POOL2
ASA-1(config-network-object)# **subnet 192.168.20.0 255.255.255.0**
ASA-1(config-network-object)# **exit**

!Configure the Policy NAT.
!The source IP of local LAN will be translated to 192.168.10.0 when the destination is 192.168.20.0

ASA-1(config)# **nat (inside,outside) source static** LOCAL-LAN NAT-POOL1 **destination static** NAT-POOL2 DEST-LAN

!Configure the Destination NAT.
!Outbound packets going to 192.168.20.0 will have their Destination IP changed to 192.168.1.0

ASA-1(config)# **object network** DEST-LAN
ASA-1(config-network-object)# **nat (outside,inside) static** NAT-POOL2

*Notice that we have **(outside,inside)** on the statement above. This creates a destination NAT.*

!Configure VPN Interesting Traffic.
ASA-1(config)# **access-list VPN-ACL extended permit ip 192.168.10.0 255.255.255.0 192.168.1.0 255.255.255.0**

Look carefully at the VPN interesting traffic above. The NAT translation goes like that:

- Assume a host in LAN-1 with source IP 192.168.1.23 wants to access destination host 192.168.1.34 in LAN-2 (remember that we have same subnets in the two LANs). The source host must access 192.168.20.34 as destination host.
- ASA will first do source NAT translation and change 192.168.1.23 into 192.168.10.23.
- Then, ASA will do also destination NAT translation and change the destination IP 192.168.20.34 into 192.168.1.34.
- The VPN interesting traffic will therefore be between 192.168.10.23 and 192.168.1.34. That's why the VPN-ACL is from source 192.168.10.0 to destination 192.168.1.0. Note that NAT is performed first in the ASA and the resulting traffic is then inserted into the VPN tunnel.

Let's see also ASA-2 configuration:

ASA 2:

!Create the required network objects which will be used in NAT
ASA-2(config)# object network obj-local
ASA-2(config-network-object)# subnet 192.168.1.0 255.255.255.0 ← LAN2
ASA-2(config-network-object)# exit

ASA-2(config)# object network DEST-LAN
ASA-2(config-network-object)# subnet 192.168.10.0 255.255.255.0 ← LAN1 mapped
ASA-2(config-network-object)# exit

!Configure the normal NAT exemption as we did in previous traditional VPN configurations.
ASA-2(config)# nat (inside,outside) source static obj-local obj-local destination static DEST-LAN DEST-LAN

!Configure VPN Interesting Traffic.
ASA-2(config)# access-list VPN-ACL extended permit ip 192.168.1.0 255.255.255.0 192.168.10.0 255.255.255.0

VERIFICATION

Let's see some output from the first ASA-1 device:

ASA1# show crypto ipsec sa

```
interface: outside
  Crypto map tag: VPNMAP, seq num: 10, local addr: 20.20.20.2

    access-list VPN-ACL extended permit ip 192.168.10.0 255.255.255.0 192.168.1.0 255.255.255.0
    local ident (addr/mask/prot/port): (192.168.10.0/255.255.255.0/0/0)
    remote ident (addr/mask/prot/port): (192.168.1.0/255.255.255.0/0/0)
    current_peer: 30.30.30.2

    #pkts encaps: 16, #pkts encrypt: 16, #pkts digest: 16
    #pkts decaps: 16, #pkts decrypt: 16, #pkts verify: 16
    #pkts compressed: 0, #pkts decompressed: 0
```

The above output shows that traffic is encrypted between 192.168.10.0 and 192.168.1.0

ASA1# sh xlate

```
3 in use, 4 most used
Flags: D - DNS, i - dynamic, r - portmap, s - static, I - identity, T - twice
NAT from inside:192.168.1.0/24 to outside:192.168.10.0/24
    flags sT idle 0:11:05 timeout 0:00:00
NAT from outside:192.168.1.0/24 to inside:192.168.20.0/24
    flags sT idle 0:11:05 timeout 0:00:00
NAT from outside:192.168.1.0/24 to inside:192.168.20.0/24
    flags s idle 1:01:12 timeout 0:00:00
```

The above output shows the static NAT mappings.

You can find a complete configuration of the scenario above in Chapter 4, Section 4.2.8.

3.1.9 Site-to-Site IKEv2 IPSEC VPN

All the IPSEC implementations described above are based on the legacy IKEv1 IPSEC. In this section we will see a simple site-to-site VPN network using the new IKEv2 IPSEC between two ASA firewalls.

The IKEv2 functionality for site-to-site is designed in-line with the existing IKEv1 implementation and it utilizes the existing configuration where appropriate and augments with IKEv2 specific configuration as necessary to allow independent control of each protocol. It provides you the same functionality that we have discussed in SITE-TO-SITE IPSEC VPN (using IKEv1) but with a few configuration differences.

In the following scenario, we will describe how IKEv2 will be used to establish a VPN tunnel between ASA-1 and ASA-2 and this will help PC 192.168.10.1 to talk to a remote Host 192.168.11.1. To make the scenario more interesting and useful, we will actually have both IKEv1 and IKEv2 configured on the ASA devices. We will use the diagram below for our scenario:

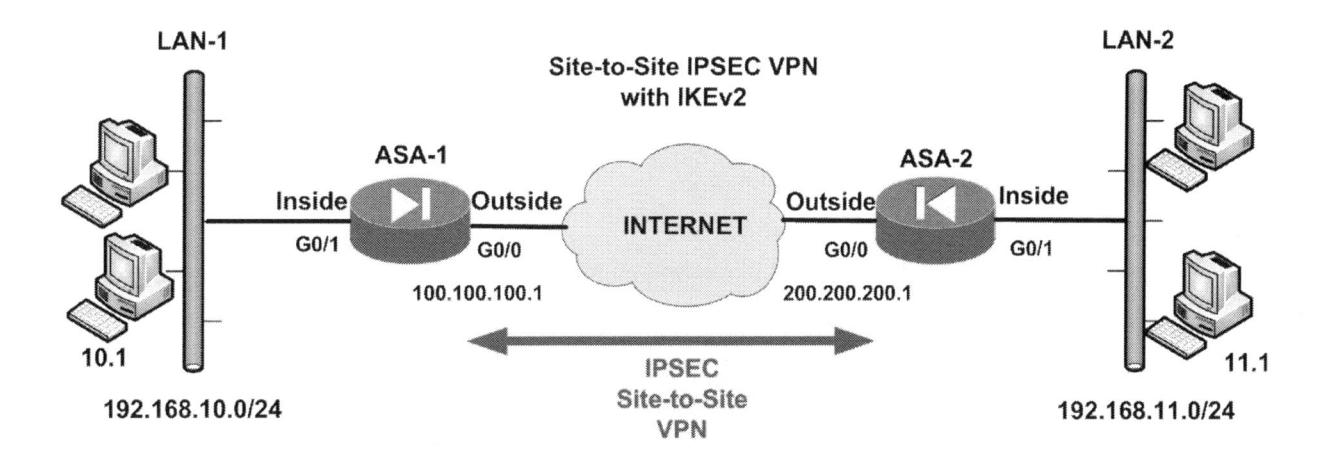

A summary of the steps required is shown on the list below:

1. Configure the ASA's :
   - We assume that Interface addresses and routing is configured already.
   - Configure Interesting Traffic to be encrypted.
   - Configure IKEv2 policies and IPSEC proposals
   - Configure IKEv1 policies and transform-sets
   - Configure Crypto map with both IKEv1 and IKEv2 IPsec policies
   - Allow IKEv2 as a vpn-tunnel-protocol in the group-policy
   - IPsec L2L tunnel-group with pre-shared-keys configured (both IKEv1 and IKEv2) under ipsec-attributes. Configure them to be different in each direction for IKEv2 to illustrate asymmetric authentication behavior.
   - Enable both IKEv1 and IKEv2 on the outside interfaces
2. Configure the workstations.
3. Send traffic across and bring the tunnel up.

Let's now see the actual configuration on the ASA Firewalls.

Step1: Configure Interesting Traffic to be encrypted

Just like the IKEv1 site-to-site VPN examples before, we need to define which traffic we want to pass through the VPN tunnel (encrypted) between LAN1 and LAN2. We can allow the whole subnet or only specific hosts. In our example, only traffic between 192.168.10.1 and 192.168.11.1 will pass through the VPN tunnel.

<u>ASA 1:</u>
ASA-1(config)# access-list LAN1-to-LAN2 extended permit ip host 192.168.10.1 host 192.168.11.1

<u>ASA 2:</u>
ASA-2(config)# access-list LAN2-to-LAN1 extended permit ip host 192.168.11.1 host 192.168.10.1

NAT Exclusion for VPN Traffic

If you are using NAT on the firewall (which is very common), you must exclude the VPN interesting traffic above from the NAT operation.

ASA 1:
ASA-1(config)# object network obj-local
ASA-1(config-network-object)# host 192.168.10.1
ASA-1(config-network-object)# exit

ASA-1(config)# object network obj-remote
ASA-1(config-network-object)# host 192.168.11.1
ASA-1(config-network-object)# exit

ASA-1(config)# nat (inside,outside) 1 source static obj-local obj-local destination static obj-remote obj-remote

ASA 2:
ASA-2(config)# object network obj-local
ASA-2(config-network-object)# host 192.168.11.1
ASA-2(config-network-object)# exit

ASA-2(config)# object network obj-remote
ASA-2(config-network-object)# host 192.168.10.1
ASA-2(config-network-object)# exit

ASA-2(config)# nat (inside,outside) 1 source static obj-local obj-local destination static obj-remote obj-remote

Step2: Configure IKEv2 Policy (similar to Phase1 in IKEv1)

Like the older IKEv1 model, we need to configure an IKEv2 policy which is similar to the **Phase1** stage we have described in IKEv1 site-to-site VPN scenario. In this policy, we can have multiple encryption and integrity protocols under the same policy. This is because IKEv2 sends across a single proposal containing multiple ciphers, compared to IKEv1 in which multiple policies must be configured if we have multiple encryption and integrity proposals.

ASA 1:
ASA-1(config)# crypto ikev2 policy 1
ASA-1(config-ikev2-policy)# encryption aes 3des ← Notice we have 2 ciphers
ASA-1(config-ikev2-policy)# integrity sha md5 ← Notice we have 2 integrity algorithms
ASA-1(config-ikev2-policy)# group 2 ← Diffie-Hellman group
ASA-1(config-ikev2-policy)# prf sha ← Pseudo Random Function Algorithm
ASA-1(config-ikev2-policy)# lifetime seconds 86400
ASA-1(config-ikev2-policy)# exit

ASA 2:
ASA-2(config)# crypto ikev2 policy 1
ASA-2(config-ikev2-policy)# encryption aes 3des ← Notice we have 2 ciphers
ASA-2(config-ikev2-policy)# integrity sha md5 ← Notice we have 2 integrity algorithms
ASA-2(config-ikev2-policy)# group 2 ← Diffie-Hellman group
ASA-2(config-ikev2-policy)# prf sha ← Pseudo Random Function Algorithm
ASA-2(config-ikev2-policy)# lifetime seconds 86400
ASA-2(config-ikev2-policy)# exit

NOTE:
PRF is the Pseudo Random Function algorithm which is same as the integrity algorithm. It is not mandatory. You must configure at least one encryption algorithm, one integrity algorithm, and one DH group for the proposal to be considered complete.

Step3: Configure IKEv2 IPSEC Proposal (similar to transform-set in IKEv1)

This is similar to the Phase2 stage we had in IKEv1 case where we have configured a "**transform set**". The "**ipsec-proposal**" in IKEv2 is the same as the "**transform-set**" we had in IKEv1.

The IPSEc security parameters in this step will be used to protect the data and messages within the tunnel.

ASA 1:
ASA-1(config)# crypto ipsec ikev2 ipsec-proposal IKEv2-AES-SHA
ASA-1(config-ipsec-proposal)# protocol esp encryption aes
ASA-1(config-ipsec-proposal)# protocol esp integrity sha-1
ASA-1(config-ipsec-proposal)# exit

ASA 2:
ASA-2(config)# crypto ipsec ikev2 ipsec-proposal IKEv2-AES-SHA
ASA-2(config-ipsec-proposal)# protocol esp encryption aes
ASA-2(config-ipsec-proposal)# protocol esp integrity sha-1
ASA-2(config-ipsec-proposal)# exit

Step4: Configure IKEv1 Policies and Transform Sets

On the same ASA device we can have both IKEv1 and IKEv2 configured. If IKEv2 VPN is not successfully established between the two ASA firewalls, they can revert back to IKEv1.
Here we set-up IKEv1 Policies and Transform Sets (as we have seen in previous section for the IKEv1 site-to-site VPN).

ASA 1:

!Configure the Phase1 Policy
ASA-1(config)# crypto ikev1 policy 10
ASA-1(config-ikev1-policy)# authentication pre-share ← Use pre-shared key for auth
ASA-1(config-ikev1-policy)# encryption aes ← Use AES encryption
ASA-1(config-ikev1-policy)# hash sha ← Use SHA for hashing
ASA-1(config-ikev1-policy)# group 2 ← Diffie-Hellman Group 2
ASA-1(config-ikev1-policy)# lifetime 86400 ← Lifetime of SA is 3600 seconds
ASA-1(config-ikev1-policy)# exit
ASA-1(config)# crypto isakmp identity address

!Configure the Phase2 Transform Set
ASA-1(config)# crypto ipsec ikev1 transform-set IKEv1-AES-SHA esp-aes esp-sha-hmac

ASA 2:

!Configure the Phase1 Policy
ASA-2(config)# crypto ikev1 policy 10
ASA-2(config-ikev1-policy)# authentication pre-share ← Use pre-shared key for auth
ASA-2(config-ikev1-policy)# encryption aes ← Use AES encryption
ASA-2(config-ikev1-policy)# hash sha ← Use SHA for hashing
ASA-2(config-ikev1-policy)# group 2 ← Diffie-Hellman Group 2
ASA-2(config-ikev1-policy)# lifetime 86400 ← Lifetime of SA is 3600 seconds
ASA-2(config-ikev1-policy)# exit
ASA-2(config)# crypto isakmp identity address

!Configure the Phase2 Transform Set
ASA-2(config)# crypto ipsec ikev1 transform-set IKEv1-AES-SHA esp-aes esp-sha-hmac

Step5: Configure a Group Policy to allow both IKEv1 and IKEv2

ASA 1:

ASA-1(config)# group-policy GroupPolicy1 internal
ASA-1(config)# group-policy GroupPolicy1 attributes
ASA-1(config-group-policy)# vpn-tunnel-protocol ikev2 ikev1 ← allow both IKEv2 IKEv1
ASA-1(config-group-policy)# exit

ASA 2:

ASA-2(config)# group-policy GroupPolicy1 internal
ASA-2(config)# group-policy GroupPolicy1 attributes
ASA-2(config-group-policy)# vpn-tunnel-protocol ikev2 ikev1 ← allow both IKEv2 IKEv1
ASA-2(config-group-policy)# exit

Step6: Configure Crypto Maps with both IKEv1 and IKEv2 IPSEC Profiles

The crypto map combines the previously created encryption algorithms, the remote peer, and the phase 2 policy into a single crypto map. Notice that we have both IKEv1 and IKEv2 IPSEC profiles attached on the same crypto map.

ASA 1:

ASA-1(config)# crypto map outside_map 1 match address LAN1-to-LAN2
ASA-1(config)# crypto map outside_map 1 set peer 200.200.200.1
ASA-1(config)# crypto map outside_map 1 set ikev1 transform-set IKEv1-AES-SHA
ASA-1(config)# crypto map outside_map 1 set ikev2 ipsec-proposal IKEv2-AES-SHA
ASA-1(config)# crypto map outside_map interface outside

ASA 2:

ASA-2(config)# crypto map outside_map 1 match address LAN2-to-LAN1
ASA-2(config)# crypto map outside_map 1 set peer 100.100.100.1
ASA-2(config)# crypto map outside_map 1 set ikev1 transform-set IKEv1-AES-SHA
ASA-2(config)# crypto map outside_map 1 set ikev2 ipsec-proposal IKEv2-AES-SHA
ASA-2(config)# crypto map outside_map interface outside

Step7: Configure Crypto Maps with both IKEv1 and IKEv2 IPSEC Profiles

At this point, we will create the tunnel group. Just like IKEv1, the preshared key (or other authentication method) is defined here. However, IKEv2 allows you to use different authentication methods for both local and remote authentication.

ASA 1:

ASA-1(config)# tunnel-group 200.200.200.1 type ipsec-l2l
ASA-1(config)# tunnel-group 200.200.200.1 general-attributes
ASA-1(config-tunnel-general)# default-group-policy GroupPolicy1←Group Policy from Step5
ASA-1(config-tunnel-general)#exit

```
ASA-1(config)# tunnel-group 200.200.200.1 ipsec-attributes
ASA-1(config-tunnel-ipsec)# ikev1 pre-shared-key cisco123
ASA-1(config-tunnel-ipsec)# ikev2 remote-authentication pre-shared-key cisco1
ASA-1(config-tunnel-ipsec)# ikev2 local-authentication pre-shared-key cisco1234
ASA-1(config-tunnel-ipsec)# exit
```

ASA 2:

```
ASA-2(config)# tunnel-group 100.100.100.1 type ipsec-l2l
ASA-2(config)# tunnel-group 100.100.100.1 general-attributes
ASA-2(config-tunnel-general)# default-group-policy GroupPolicy1 ←Group Policy from Step5
ASA-2(config-tunnel-general)#exit

ASA-2(config)# tunnel-group 100.100.100.1 ipsec-attributes
ASA-2(config-tunnel-ipsec)# ikev1 pre-shared-key cisco123
ASA-2(config-tunnel-ipsec)# ikev2 remote-authentication pre-shared-key cisco1234
ASA-2(config-tunnel-ipsec)# ikev2 local-authentication pre-shared-key cisco1
ASA-2(config-tunnel-ipsec)# exit
```

NOTE:

Please note that the pre-shared-keys are used to authenticate the remote peer in order to build a trust relationship. If you compare the configuration on ASA1 and ASA2, you will see that the pre-shared-key defined for **remote-authentication** on ASA1 is matching the pre-shared-key defined for **local authentication** on ASA2 and vice versa. This illustrates the asymmetrical authentication allowed on IKEv2.

Step8: Enable both IKEv1 and IKEv2 on outside interface

ASA 1:
```
ASA-1(config)# crypto ikev2 enable outside
ASA-1(config)# crypto ikev1 enable outside
```

ASA 2:
```
ASA-2(config)# crypto ikev2 enable outside
ASA-2(config)# crypto ikev1 enable outside
```

Step9: Verification

ASA-1# show crypto isakmp sa

> There are no IKEv1 SAs
>
> IKEv2 SAs:
>
> Session-id:1, Status:UP-ACTIVE, IKE count:1, CHILD count:1
>
> Tunnel-id   Local   Remote   Status   Role
> 9807541   100.100.100.1/500   200.200.200.1/500   READY   INITIATOR
>   Encr: AES-CBC, keysize: 128, Hash: SHA96, DH Grp:2, Auth sign: PSK, Auth verify: PSK
>   Life/Active Time: 86400/58 sec
> Child sa: local selector 192.168.10.1/0 - 192.168.10.1/65535
>   remote selector 192.168.11.1/0 - 192.168.11.1/65535
>   ESP spi in/out: 0x19e57b7b/0x5520a043

As you have seen above, the ASA firewall has established an IKEv2 Security Association (SA) with the remote peer. If you have both IKEv1 and IKEv2 on the same device, then IKEv2 is preferred.

ASA-1# show crypto ipsec sa

> interface: outside
>   Crypto map tag: outside_map, seq num: 1, local addr: 100.100.100.1
>
>   access-list LAN1-to-LAN2 extended permit ip host 192.168.10.1 host 192.168.11.1
>   [output omitted]......
>
>   #pkts encaps: 7, #pkts encrypt: 7, #pkts digest: 7
>   #pkts decaps: 7, #pkts decrypt: 7, #pkts verify: 7

There are also packets encrypted and decrypted accordingly as shown above.

3.2   SSL-Based VPN Configuration on Cisco ASA

SSL Based is the newest VPN type on ASA firewalls. It is used only for Remote Access implementations and provides flexibility and lower administration overhead since no IPSEC Client software is required to be installed manually on user's computers.

3.2.1 Anyconnect SSL Web VPN

The Anyconnect SSL VPN implementation is the most powerful option since it provides full network access to remote users. This is similar with the IPSEC VPN client software which also provides full network access remotely. The newest Anyconnect product from Cisco is called now "**Cisco Anyconnect Secure Mobility Client**". From Anyconnect Client version 3.x and above both SSL and IKEv2/IPSEC protocols are supported.

There are two Initial Installation options for AnyConnect client:
- Using clientless WebVPN portal.
- Manual installation by the user or administrator

Using the clientless Web portal, the user first connects and authenticates securely to the ASA with a web browser and the Java Anyconnect client is automatically downloaded and installed on the user's computer (the user can also click the "AnyConnect" Tab on the WebVPN portal to download the client). This necessitates that the Java client (**.pkg extension**) must be already stored on the ASA flash memory by the administrator. After the Anyconnect client is downloaded and installed upon the first connection, the user from now on can start the Anyconnect client directly from his/her computer and connect remotely without using a web browser. This is the preferred method in my opinion because it automates the distribution of the client to the remote users.

With the manual installation method, the network administrator must download the appropriate Anyconnect client software (Microsoft package or one of the other OS versions) from Cisco site and provide the file to the users for manual installation on their laptop. With this method, the user does not need to log in via clientless mode to start the SSL VPN tunnel. Instead, the users can start up the AnyConnect client manually from their desktop and provide their authentication credentials.

Let's see how to configure a Cisco ASA for Anyconnect SSL VPN based on the diagram below.

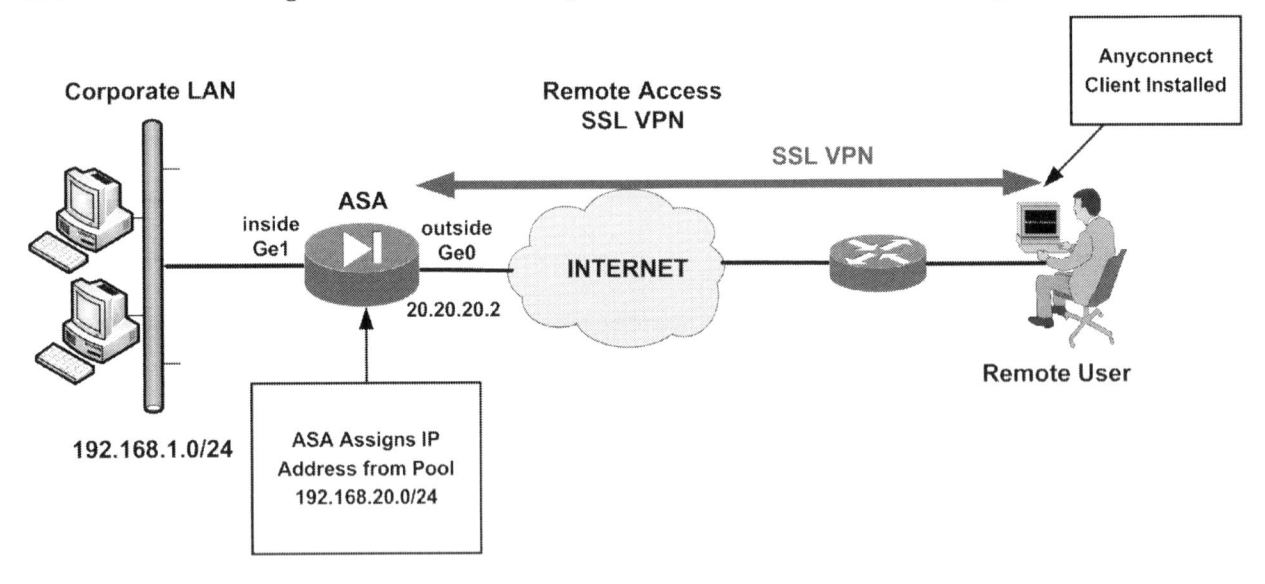

ASA:

- **STEP 1: Copy Anyconnect Software to ASA Flash**

As we've said before, we need to transfer the Anyconnect package file to the flash of the ASA. First you need to download one of the **.pkg** files from Cisco website. An example Windows client file has the format "**anyconnect-win-x.x.xxxx-k9.pkg**".

To copy the PKG file to ASA flash:
ASA# copy {tftp|ftp|scp}://[ip address]/anyconnect-win-x.x.xxxx-k9.pkg **disk0:**

Assume we have downloaded the Anyconnect client file on our computer with IP address 192.168.1.1. We will use a TFTP server on our PC to transfer the file to ASA.

ASA# copy tftp://192.168.1.1/anyconnect-win-2.4.1012-k9.pkg disk0:

Address or name of remote host [192.168.1.1]?

Source filename [anyconnect-win-2.4.1012-k9.pkg]?

Destination filename [anyconnect-win-2.4.1012-k9.pkg]?

Accessing tftp://192.168.1.1/anyconnect-win-2.4.1012-k9.pkg...!!!!!!

- **STEP 2: Tell the ASA where to find the Anyconnect software on Flash**

Then we need to identify the PKG image file on flash by telling the ASA where the image file is located. Also, enable the webvpn Anyconnect service on the outside ASA interface.

ASA# configure terminal
ASA(config)# webvpn
ASA(config-webvpn)# anyconnect image disk0:/anyconnect-win-2.4.1012-k9.pkg 1
ASA(config-webvpn)# enable outside ← enable ssl webvpn on outside interface
ASA(config-webvpn)# anyconnect enable ← enable anyconnect service
ASA(config-webvpn)# exit

Note: The number 1 at the end of the package file is the file order. It is used when you have more than one images stored on the ASA flash (e.g Anyconnect client images for Windows and MAC).

- **STEP 3: Configure VPN Pool to assign IP addresses**

Create an IP address pool from which the ASA will assign addresses to remote users. From the diagram above we see that after the remote user gets authenticated, the ASA assigns an IP address to the remote user from a predefined pool 192.168.20.0/24

ASA(config)# ip local pool vpnpool 192.168.20.1-192.168.20.254 mask 255.255.255.0

- **STEP 4: Configure NAT Exemption**

Create a NAT exemption for traffic between the corporate LAN network behind the ASA (192.168.1.0/24) and the remote user's address pool (192.168.20.0/24).

ASA(config)# object network obj-local
ASA(config-network-object)# subnet 192.168.1.0 255.255.255.0
ASA(config-network-object)# exit

ASA(config)# object network obj-vpnpool
ASA(config-network-object)# subnet 192.168.20.0 255.255.255.0
ASA(config-network-object)# exit

ASA(config)# nat (inside,outside) source static obj-local obj-local destination static obj-vpnpool obj-vpnpool no-proxy-arp route-lookup

- **STEP 5: Configure Split-Tunneling (Optional)**

Similar with the IPSEC VPN client configuration, if you want to allow users to access the Internet and also access the Corporate LAN network at the same time, you must configure a Split-Tunnel Access Control List.

ASA(config)# access-list split-tunnel standard permit 192.168.1.0 255.255.255.0

Traffic from the remote users towards the network specified in the split-tunnel ACL (192.168.1.0/24) will pass through the SSL VPN tunnel. All other traffic from the remote user will go to the Internet.

- **STEP 6: Configure VPN Group Policy**

Create a Group Policy for the AnyConnect WebVPN users. The Group Policy allows you to separate different remote access users into groups with different attributes. The Group Policy attributes that can be configured include DNS server addresses, split-tunneling settings, how the client will be downloaded (automatically or after prompting the user), if the client software will remain permanently on the user's computer etc.

The command format is as following:

ASA(config)# group-policy *"policy name"* **internal**
ASA(config)# group-policy *"policy name"* **attributes**
ASA(config-group-policy)# vpn-tunnel-protocol {[ikev1] [ikev2][l2tp-ipsec][ssl-client]}
ASA(config-group-policy)# split-tunnel-policy {tunnelspecified | tunnelall}
ASA(config-group-policy)# split-tunnel-network-list value *"acl-for-split-tunnel"*
ASA(config-group-policy)# webvpn
ASA(config-group-webvpn)# anyconnect keep-installer {installed | none}
ASA(config-group-webvpn)# anyconnect ask {none | enable [default {webvpn | anyconnect} timeout *value*]**}**

Let's clarify some of the Group Policy commands shown above:

vpn-tunnel-protocol {[ikev1] [ikev2][l2tp-ipsec][ssl-client]} ← Select the type of VPN tunnel protocol. For SSL VPN you must select "ssl-client"

split-tunnel-policy {tunnelspecified | tunnelall} ← Specify whether only selected traffic will pass through the tunnel ("tunnelspecified") or whether ALL remote traffic will pass through the tunnel ("tunnelall").

split-tunnel-network-list value *"acl-for-split-tunnel"* ← Specify the Access List for split-tunnel (see Step 5 above)

anyconnect keep-installer {installed | none} ← "installed" means that the client remains installed permanently on the user's computer even after disconnection. The default is that the client gets uninstalled after the user disconnects from the Anyconnect session.

anyconnect ask {none | enable [default {webvpn | anyconnect } timeout *value*]} ← This command has to do with how AnyConnect client will be downloaded to user's computer.

- **anyconnect ask none default webvpn** ← The ASA immediately displays the WebPortal. This is the default configuration.
- **anyconnect ask none default anyconnect** ← Download the AnyConnect client automatically.
- **anyconnect ask enable default anyconnect timeout 20** ← The user will get a prompt to install the AnyConnect client. If nothing is done within 20 seconds, the client will be downloaded and installed automatically.

anyconnect dpd-interval {[gateway {*seconds* / none}] / [client {*seconds* / none}]} ← This enables Dead Peer Detection (DPD) mechanism which ensures that the ASA (gateway) or the client can quickly detect a condition where the peer is not responding and the connection has failed.

Let's see the actual configuration commands of group-policy for our specific scenario:

EXAMPLE:

ASA(config)# **group-policy SSLVPNpolicy internal**

ASA(config)# **group-policy SSLVPNpolicy attributes**

ASA(config-group-policy)# **vpn-tunnel-protocol ssl-client**

ASA(config-group-policy)# **split-tunnel-policy tunnelspecified**

ASA(config-group-policy)# **split-tunnel-network-list value split-tunnel**

ASA(config-group-policy)# **webvpn**

ASA(config-group-webvpn)# **anyconnect keep-installer installed**

ASA(config-group-webvpn)# **anyconnect ask enable default anyconnect timeout 10**

ASA(config-group-webvpn)# **anyconnect dpd-interval client 20** ← The client will check for Dead Peer Detection every 20 seconds.

- **STEP 7: Configure a Tunnel Group**

Create a <u>Tunnel Group.</u> The tunnel group must incorporate the Group Policy configured above. It also binds the Group Policy with the IP address pool that we have already configured for remote users.

The command format is as following:

ASA(config)# **tunnel-group** *"tunnel name"* **type remote-access**

ASA(config)# **tunnel-group** *"tunnel name"* **general-attributes**

ASA(config-tunnel-general)# **default-group-policy** *"group policy name"* ←Assign the Group Policy configured in Step6 above.

ASA(config-tunnel-general)# **address-pool** *"IP Pool for VPN"* ← Assign the IP address pool configured in Step3 above.

ASA(config-tunnel-general)# **exit**

ASA(config)# **tunnel-group** *"tunnel name"* **webvpn-attributes**

ASA(config-tunnel-webvpn)# **group-alias** *"group_name_alias"* **enable** ← Create an alias name for the tunnel group which will be listed on the log-in screen of the Anyconnect client.

ASA(config-tunnel-webvpn)# **exit**

ASA(config)# **webvpn**

ASA(config-webvpn)# tunnel-group-list enable ← Enable the listing of the alias name on the log-in screen of the AnyConnect client.

Let's see the actual configuration commands of tunnel-group for our specific scenario:
EXAMPLE:
ASA(config)# tunnel-group SSLVPNprofile type remote-access
ASA(config)# tunnel-group SSLVPNprofile general-attributes
ASA(config-tunnel-general)# default-group-policy SSLVPNpolicy
ASA(config-tunnel-general)# address-pool vpnpool
ASA(config-tunnel-general)# exit
ASA(config)# tunnel-group SSLVPNprofile webvpn-attributes
ASA(config-tunnel-webvpn)# group-alias SSL_USERS enable ← This name will be shown to the log-in screen of Anyconnect.
ASA(config-tunnel-webvpn)# exit
ASA(config)# webvpn
ASA(config-webvpn)# tunnel-group-list enable ← Allow users to select which tunnel group to connect (useful if you have multiple tunnel groups)

- **STEP 8: Configure Local User(s) for Authentication**

Create a local user on ASA which will be used for AnyConnect authentication. This user will be allowed to have remote network access.

ASA(config)# username sslvpnuser password test123
ASA(config)# username sslvpnuser attributes ← OPTIONAL
ASA(config-username)# service-type remote-access ← OPTIONAL

- **STEP 9: Verification**

Let's see some screenshots and ASA log output for verification.

1. Connect to ASA on its public outside address: **https://[outside ASA Address]**
You might have to accept some certificate messages. Then, you will get the following log-in screen:

2. Enter your username and password (sslvpnuser). Also, choose your respective group from the drop down list as shown. Note that the group name in the drop down is the group-alias name configured in Step7 (**SSL_USERS**). After successful authentication you can see that a VPN Session is established with 20.20.20.2 (Public IP of ASA).

3. Clicking on the "Statistics" Tab you can see various important stats and parameters as shown below:

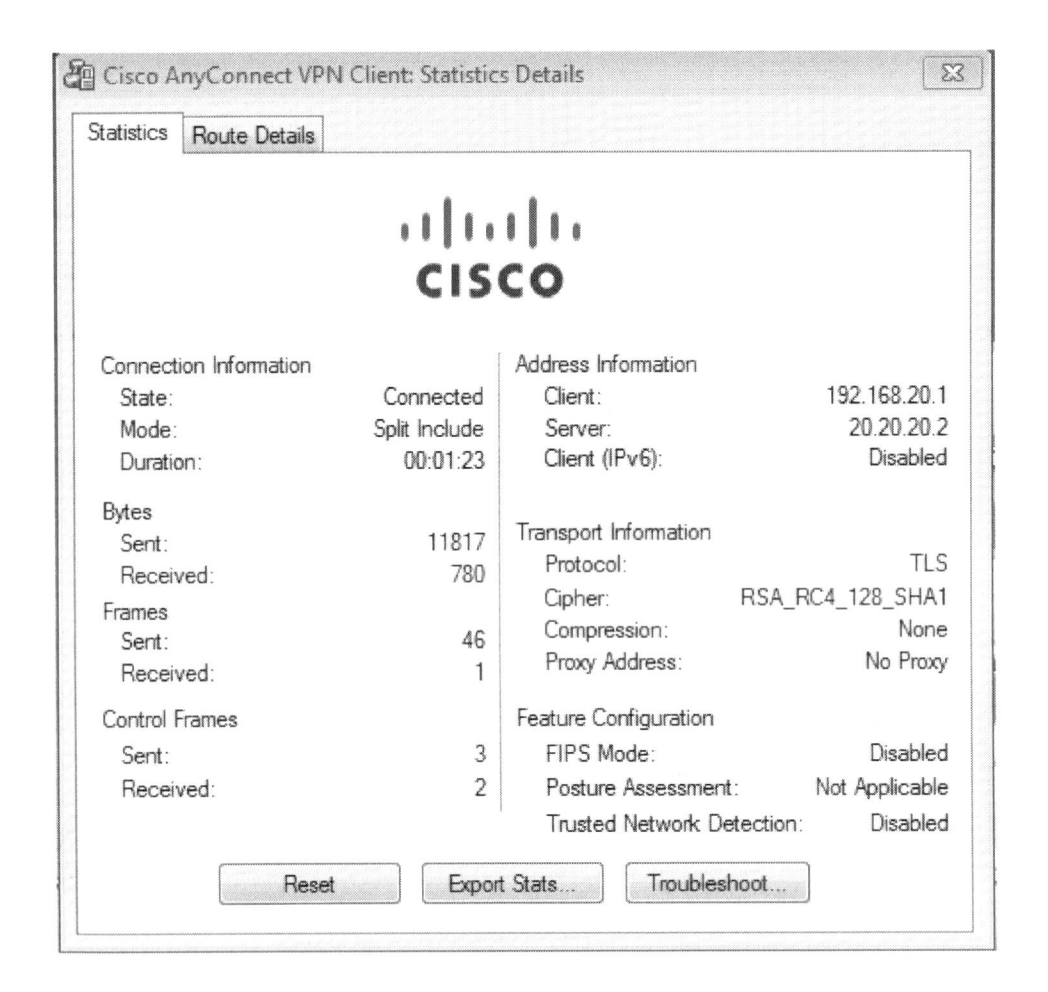

The statistics tab above shows that the Anyconnect remote user received an IP address 192.168.20.1 from the ASA.

4. Let's see also some output from the ASA:
ASA# show webvpn anyconnect

1. disk0:/anyconnect-win-2.4.1012-k9.pkg 1 dyn-regex=/Windows NT/
CISCO STC win2k+
2,4,1012
Thu 12/17/2009 15:47:55.45

1 AnyConnect Client(s) installed

ASA# show ip local pool vpnpool

Begin	End	Mask	Free	Held	In use
192.168.20.1	192.168.20.254	255.255.255.0	253	0	1

In Use Addresses:
192.168.20.1

The above verifies that there is one Anyconnect user connected who received an IP 192.168.20.1

You can find a complete configuration of the scenario above in Chapter 4, Section 4.2.9.

3.3 VPN Authentication using External Server

In all of our scenarios we have seen so far, the authentication of remote access users was implemented using local device username/password credentials. That is, local user credentials were created on the device (ASA or Router) which were used to authenticate remote access users (either for IPSEC VPN or for Anyconnect SSL VPN). However, if you have a large number of remote users, it's not manageable to create local device credentials for all of them. The best option for such a case is to use an external authentication server which will hold all remote users' credentials for authentication. We will see three popular options for External Server authentication: Using **Microsoft Active Directory**, using a **AAA Radius/Tacacs** server (such as as the Cisco Secure ACS Server), and finally using an **RSA Server** for two-factor authentication.

3.3.1 VPN Authentication using Microsoft Active Directory

In this section we will describe how to implement user VPN authentication on ASA devices via a Microsoft Active Directory. This is very useful in cases where there are a large number of remote users who require VPN access to network resources via an ASA firewall, and these users already have Active Directory accounts. Therefore, administrators won't need to create and maintain extra account credentials on the ASA device.

With the addition of LDAP support on Cisco ASA firewalls, it is possible now to use a Microsoft Active Directory (AD) server to authenticate remote access users. As we know, AD supports the LDAP protocol.

There are two general steps to configure AD authentication of remote access users on Cisco ASA:

1. First configure a AAA server group which will be using the LDAP protocol. Under this group, define the parameters of the Active Directory server (IP address, distinguished names, AD login username/password etc).
2. After proper configuration of the AAA server group above, assign this group to the desired connection profile ("**Tunnel Group**") of the remote access users.

Let's see the steps above in more details. We will be using the network diagram below:

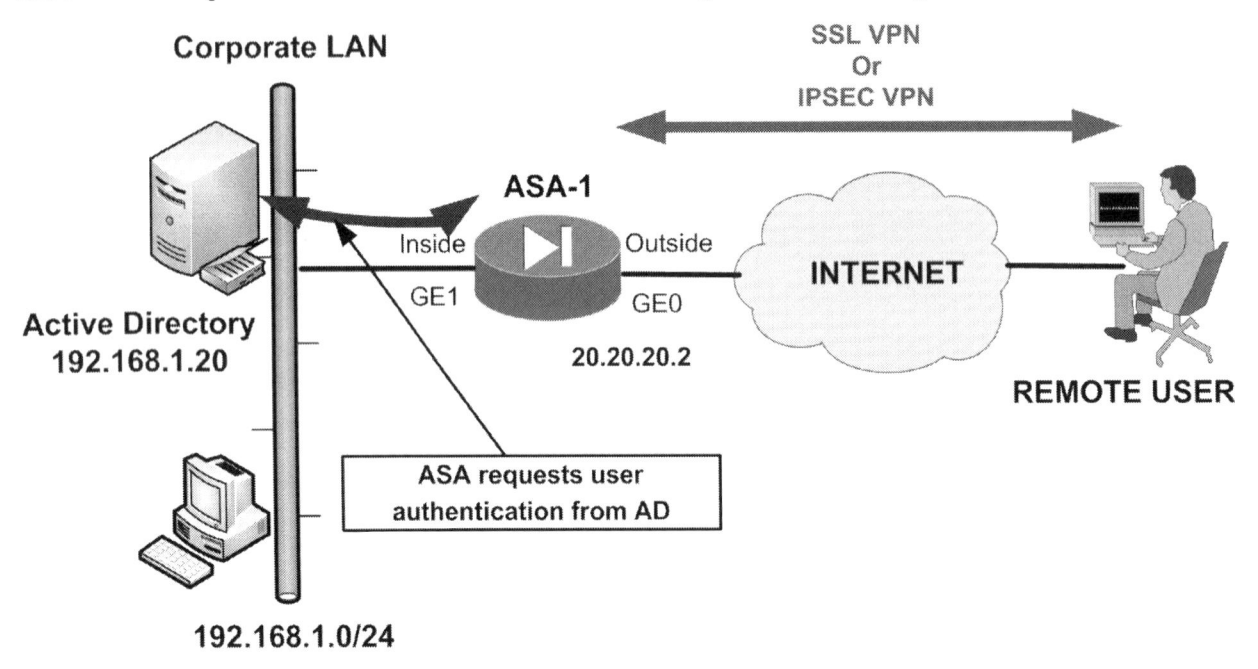

Assume we have remote access users which are connected either via the traditional IPSEC VPN client or via the Anyconnect SSL VPN method. An internal Active Directory Server (192.168.1.20) will be used by the ASA device to send the authentication requests from remote users.

- **STEP 1: Configure AAA Server Group and LDAP parameters**

ASA-1(config)# **aaa-server AD-SERVER protocol ldap** ← The name "AD-SERVER" will be used later under a Tunnel Group profile. This server uses the "ldap" protocol.
ASA-1(config-aaa-server-group)# **exit**
ASA-1(config)# **aaa-server AD-SERVER (inside) host 192.168.1.20** ← The specific "AD-SERVER" is reachable via the "inside" interface on IP 192.168.1.20
ASA-1(config-aaa-server-host)# **server-type microsoft** ← This AAA server is "Microsoft"
ASA-1(config-aaa-server-host)# **ldap-base-dn dc=mycompany, dc=com** ← See below
ASA-1(config-aaa-server-host)# **ldap-login-dn cn=admin, cn=users, dc=mycompany, dc=com** ← See below

ASA-1(config-aaa-server-host)# **ldap-login-password** *cisco123* ← See below
ASA-1(config-aaa-server-host)# **ldap-naming-attribute sAMAccountName** ← See below
ASA-1(config-aaa-server-host)# **ldap-scope subtree** ← See below

The configuration parameters in red above are explained below:

- **ldap-base-dn** : Specifies the location in the LDAP hierarchy where the server should begin searching when it receives an authentication request from ASA.

- **ldap-login-dn** : Specifies the Distinguished Name (DN) for the admin account or any account on the Active Directory which has the privileges to login, search and retrieve account information from the AD. Here we used the username "**admin**" as an example. You must use a proper username which has enough privileges to be able to search/read/lookup users in the LDAP server.

- **ldap-login-password** : Specifies the password of the "admin" account used in "ldap-login-dn" parameter above.

- **ldap-naming-attribute** : Specifies the Relative Distinguished Name (DN) attribute that uniquely identifies an entry on the LDAP server. **sAMAccountName** is the default attribute in the Microsoft Active Directory.

- **ldap-scope** : This specifies whether ASA will look at the base DN level or go below the Base DN level to search for the user accounts. In our case we want to go below the Base DN level, so we use the "**subtree**" value.

The above concludes the mandatory configuration parameters required for properly specifying an Active Directory (LDAP) server to be used by the ASA for user authentication. Next we will see how to apply the AAA Server Group above to a VPN connection profile (Tunnel-Group) in order to be used for authentication.

- **STEP 2: Assign the above AAA Server Group to a VPN Tunnel-Group**

When we discussed the remote access scenarios for both IPSEC VPN and Anyconnect VPN (sections 3.1.4 and 3.2.1) we have seen that one of the required elements to configure is a "**tunnel-group**". In order to use the AAA Server Group configured above for authentication via AD, we must assign it under the Tunnel-Group profile.

ASA-1(config)# tunnel-group remotevpn type remote-access
ASA-1(config)# tunnel-group remotevpn general-attributes
ASA-1(config-tunnel-general)# authentication-server-group AD-SERVER ← Assign the AAA Server Group from Step 1 above. Anyone using the "remotevpn" group for remote access, will be authenticated via the "AD-SERVER" using Active Directory.

3.3.2 VPN Authentication using RADIUS or TACACS

Another popular method for authentication of remote VPN users is with an external AAA Server which uses the RADIUS or TACACS protocol. For example, the Cisco Secure Access Control System (CS-ACS) supports both RADIUS and TACACS+ protocols, so you can use it in conjunction with a Cisco ASA to authenticate remote access VPN users. Furthermore, the Cisco ACS server can communicate with a two-factor authentication server (such as RSA) to provide two-factor authentication of remote access VPN users (e.g providing One-Time-Passwords with a token), as we will see later.

The general configuration steps are the same as with Active Directory above. You need to define a AAA Server Group and then attach it to a VPN connection profile ("Tunnel Group"). Let's see the configuration steps based on the diagram below:

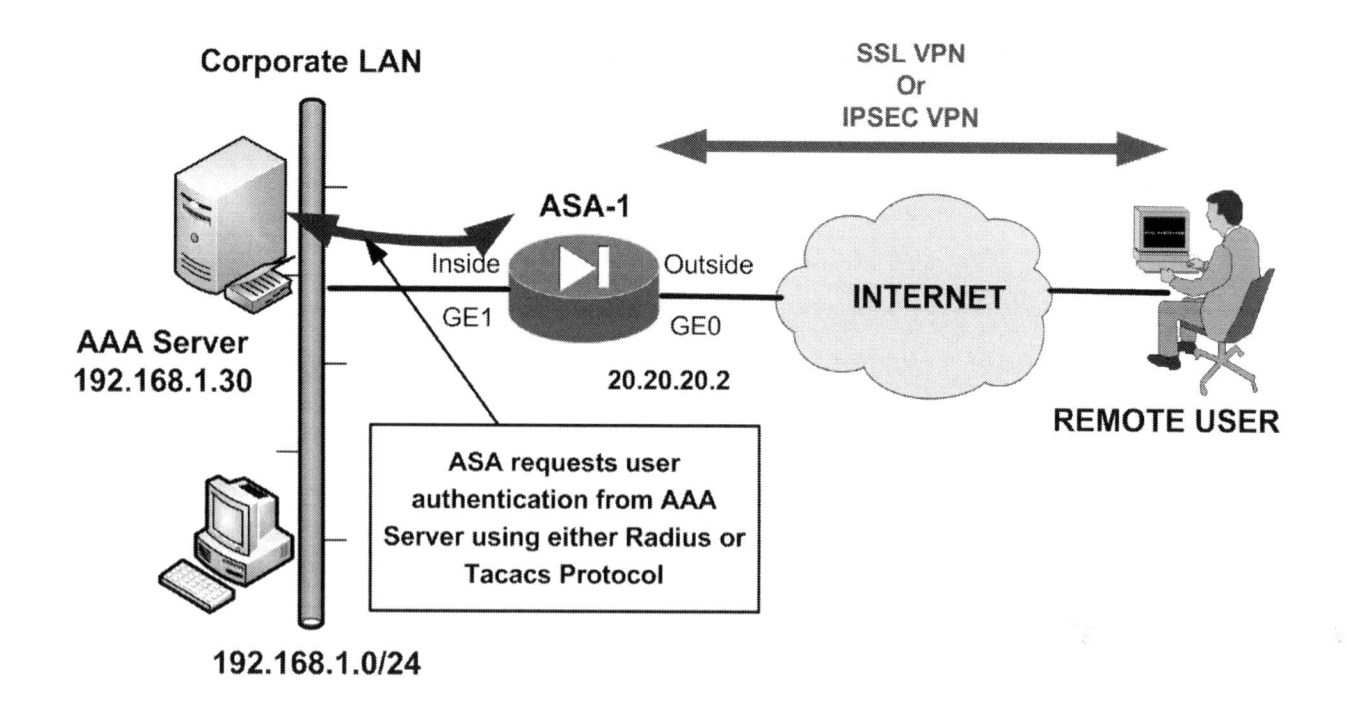

- STEP 1: Configure AAA Server Group

ASA-1(config)# aaa-server AAA-SERVER protocol [radius|tacacs+] ← The name "AAA-SERVER" will be used later under a Tunnel Group profile. This server will use either "radius" or "tacacs+" protocol.
ASA-1(config-aaa-server-group)# exit
ASA-1(config)# aaa-server AAA-SERVER (inside) host 192.168.1.30 ← The specific "AAA-SERVER" is reachable via the "inside" interface on IP 192.168.1.30
ASA-1(config-aaa-server-host)# key *strongkey* ← Authentication password between ASA and External AAA Server

- STEP 2: Assign the above AAA Server Group to a VPN Tunnel-Group

ASA-1(config)# tunnel-group remotevpn type remote-access
ASA-1(config)# tunnel-group remotevpn general-attributes
ASA-1(config-tunnel-general)# authentication-server-group AAA-SERVER ← Assign the AAA Server Group from Step 1 above. Anyone using the "remotevpn" group for remote access, will be authenticated via the "AAA-SERVER".

3.3.3 VPN Authentication using RSA

RSA is popular for providing two-factor authentication for remote access users. Using either a hardware or software token on the user side, the RSA server can issue One-Time Passwords to remote users. It's not in the scope of this book to describe the details of configuring the RSA server itself. However, we will see the configuration on the ASA to communicate with an RSA server for authentication.

There are two authentication options to use with ASA and RSA.
1. ASA communicates with a RADIUS server (usually a Cisco Secure ACS Server) for authentication (just like Section 3.3.2 above) and the RADIUS communicates with the RSA server for One-Time Passwords.
2. ASA communicates with RSA Server directly. This is what we will see below.

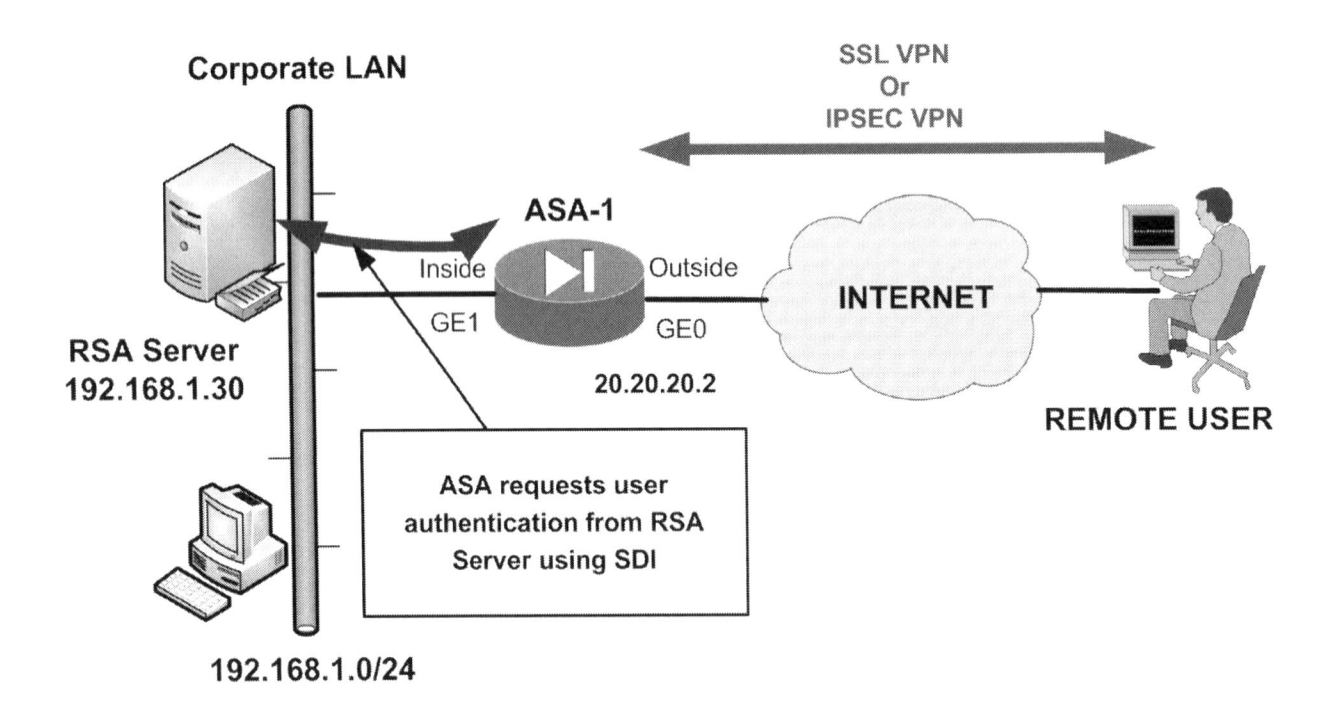

- STEP 1: Configure AAA Server Group

ASA-1(config)# aaa-server RSA-SERVER protocol sdi ← Use "SDI" as protocol
ASA-1(config-aaa-server-group)# exit
ASA-1(config)# aaa-server RSA-SERVER (inside) host 192.168.1.30 ← The specific "RSA-SERVER" is reachable via the "inside" interface on IP 192.168.1.30

- **STEP 2: Assign the above AAA Server Group to a VPN Tunnel-Group**

ASA-1(config)# tunnel-group remotevpn type remote-access
ASA-1(config)# tunnel-group remotevpn general-attributes
ASA-1(config-tunnel-general)# authentication-server-group RSA-SERVER ← Assign the AAA Server Group from Step 1 above. Anyone using the "remotevpn" group for remote access, will be authenticated via the "RSA-SERVER".

Chapter 4   Complete Configuration Examples

In this Chapter you will find complete configurations for all the scenarios that we have discussed in previous Chapters. Having the complete configuration commands as taken from actual devices is great to help you configure VPNs in Routers and ASA firewalls from end to end.

4.1   Complete VPN Configurations on Cisco Routers

4.1.1 Site-to-Site IPSEC VPN

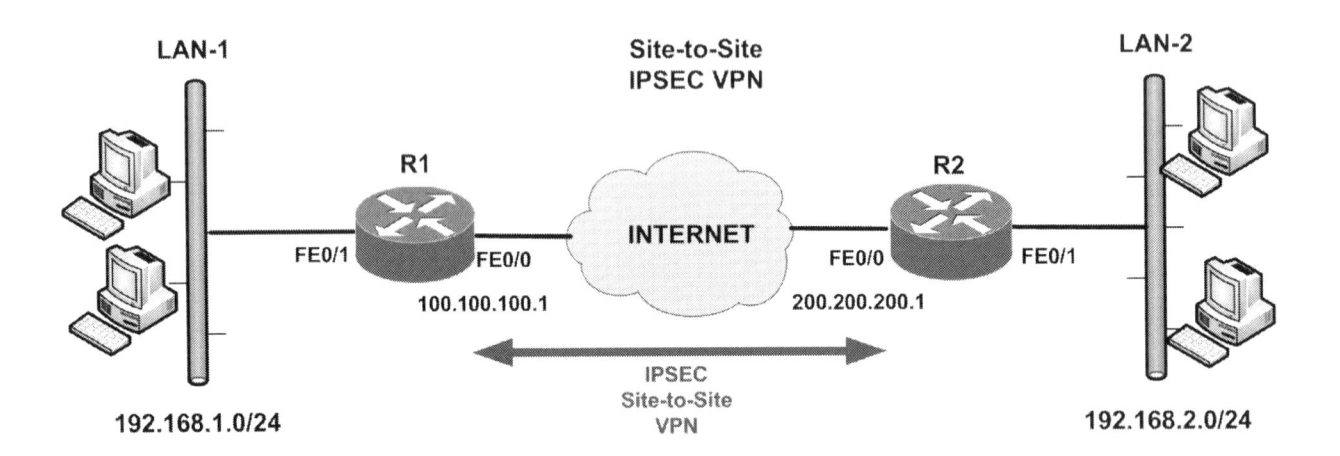

R1

```
version 12.4
service timestamps debug datetime msec
service timestamps log datetime msec
no service password-encryption
!
hostname R1
!
boot-start-marker
boot-end-marker
!
!
no aaa new-model
memory-size iomem 5
ip cef
!
!
```

```
 ip auth-proxy max-nodata-conns 3
 ip admission max-nodata-conns 3
!
crypto isakmp policy 1
 encr 3des
 hash md5
 group 2
 authentication pre-share
crypto isakmp key secretkey address 200.200.200.1
!
!
crypto ipsec transform-set TRSET esp-3des esp-md5-hmac
!
crypto map VPNMAP 10 ipsec-isakmp
 set peer 200.200.200.1
 set transform-set TRSET
 match address VPN-ACL
!
!
interface FastEthernet0/0
 ip address 100.100.100.1 255.255.255.0
 ip nat outside
 ip virtual-reassembly
 duplex auto
 speed auto
 crypto map VPNMAP
!
interface FastEthernet0/1
 ip address 192.168.1.1 255.255.255.0
 ip nat inside
 ip virtual-reassembly
 duplex auto
 speed auto
!
ip forward-protocol nd
ip route 0.0.0.0 0.0.0.0 100.100.100.2
!
no ip http server
no ip http secure-server
ip nat inside source list NAT-ACL interface FastEthernet0/0 overload
!
ip access-list extended NAT-ACL
 deny   ip 192.168.1.0 0.0.0.255 192.168.2.0 0.0.0.255
 permit ip 192.168.1.0 0.0.0.255 any
ip access-list extended VPN-ACL
 permit ip 192.168.1.0 0.0.0.255 192.168.2.0 0.0.0.255
```

```
!
control-plane
!
line con 0
line aux 0
line vty 0 4
 login
!
end
```

R2

```
version 12.4
service timestamps debug datetime msec
service timestamps log datetime msec
no service password-encryption
!
hostname R2
!
boot-start-marker
boot-end-marker
!
!
no aaa new-model
memory-size iomem 5
ip cef
!
!
ip auth-proxy max-nodata-conns 3
ip admission max-nodata-conns 3
!
crypto isakmp policy 1
 encr 3des
 hash md5
 group 2
 authentication pre-share
crypto isakmp key secretkey address 100.100.100.1
!
!
crypto ipsec transform-set TRSET esp-3des esp-md5-hmac
!
crypto map VPNMAP 10 ipsec-isakmp
 set peer 100.100.100.1
 set transform-set TRSET
```

```
  match address VPN-ACL
 !
 !
 interface FastEthernet0/0
  ip address 200.200.200.1 255.255.255.0
  ip nat outside
  ip virtual-reassembly
  duplex auto
  speed auto
  crypto map VPNMAP
 !
 interface FastEthernet0/1
  ip address 192.168.2.1 255.255.255.0
  ip nat inside
  ip virtual-reassembly
  duplex auto
  speed auto
 !
 ip forward-protocol nd
 ip route 0.0.0.0 0.0.0.0 200.200.200.2
 !
 no ip http server
 no ip http secure-server
 ip nat inside source list NAT-ACL interface FastEthernet0/0 overload
 !
 ip access-list extended NAT-ACL
  deny   ip 192.168.2.0 0.0.0.255 192.168.1.0 0.0.0.255
  permit ip 192.168.2.0 0.0.0.255 any
 !
 ip access-list extended VPN-ACL
  permit ip 192.168.2.0 0.0.0.255 192.168.1.0 0.0.0.255
 !
 control-plane
 !
 line con 0
 line aux 0
 line vty 0 4
  login
 !
 end
```

4.1.2 Site-to-Site IPSEC VPN with Dynamic IP

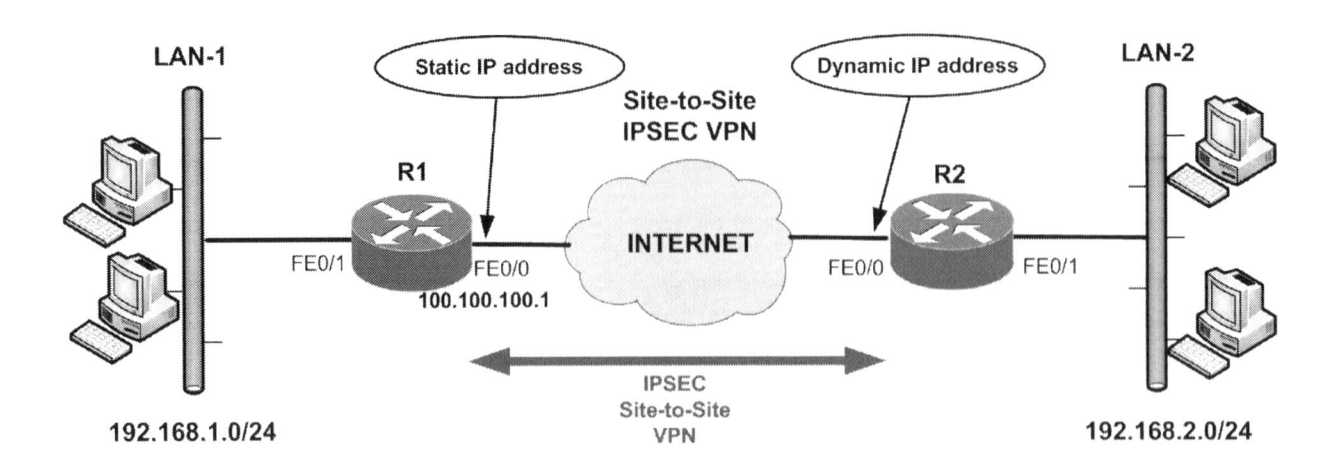

R1

```
version 12.4
service timestamps debug datetime msec
service timestamps log datetime msec
no service password-encryption
!
hostname R1
!
boot-start-marker
boot-end-marker
!
no aaa new-model
memory-size iomem 5
ip cef
!
ip auth-proxy max-nodata-conns 3
ip admission max-nodata-conns 3
!
crypto isakmp policy 10
 encr 3des
 hash md5
 authentication pre-share
 group 2
crypto isakmp key secretkey address 0.0.0.0 0.0.0.0
!
!
crypto ipsec transform-set TRSET esp-3des esp-md5-hmac
```

!
crypto dynamic-map DYNMAP 10
 set transform-set TRSET
 match address VPN-ACL
!
!
crypto map VPNMAP 10 ipsec-isakmp dynamic DYNMAP
!
!
interface FastEthernet0/0
 ip address 100.100.100.1 255.255.255.0
 ip nat outside
 ip virtual-reassembly
 speed 100
 full-duplex
 crypto map VPNMAP
!
interface FastEthernet0/1
 ip address 192.168.1.1 255.255.255.0
 ip nat inside
 ip virtual-reassembly
 duplex auto
 speed auto
!
ip forward-protocol nd
ip route 0.0.0.0 0.0.0.0 100.100.100.2
!
no ip http server
no ip http secure-server
ip nat inside source list NAT-ACL interface FastEthernet0/0 overload
!
ip access-list extended NAT-ACL
 deny   ip 192.168.1.0 0.0.0.255 192.168.2.0 0.0.0.255
 permit ip 192.168.1.0 0.0.0.255 any
ip access-list extended VPN-ACL
 permit ip 192.168.1.0 0.0.0.255 192.168.2.0 0.0.0.255
!
control-plane
!
line con 0
line aux 0
line vty 0 4
 login
!
!
End

R2

```
version 12.4
service timestamps debug datetime msec
service timestamps log datetime msec
no service password-encryption
!
hostname R2
!
boot-start-marker
boot-end-marker
!
no aaa new-model
memory-size iomem 5
ip cef
!
ip auth-proxy max-nodata-conns 3
ip admission max-nodata-conns 3
!
no ip domain lookup
!
crypto isakmp policy 10
 encr 3des
 hash md5
 authentication pre-share
 group 2
crypto isakmp key secretkey address 100.100.100.1
!
!
crypto ipsec transform-set TRSET esp-3des esp-md5-hmac
!
crypto map VPNMAP 10 ipsec-isakmp
 set peer 100.100.100.1
 set transform-set TRSET
 match address VPN-ACL
!
!
 interface FastEthernet0/0
  ip address dhcp
  ip nat outside
  ip virtual-reassembly
  speed 100
  full-duplex
  crypto map VPNMAP
 !
 interface FastEthernet0/1
```

```
 ip address 192.168.2.1 255.255.255.0
 ip nat inside
 ip virtual-reassembly
 duplex auto
 speed auto
!
ip forward-protocol nd
!
no ip http server
no ip http secure-server
ip nat inside source list NAT-ACL interface FastEthernet0/0 overload
!
ip access-list extended NAT-ACL
 deny   ip 192.168.2.0 0.0.0.255 192.168.1.0 0.0.0.255
 permit ip 192.168.2.0 0.0.0.255 any
ip access-list extended VPN-ACL
 permit ip 192.168.2.0 0.0.0.255 192.168.1.0 0.0.0.255
!
control-plane
!
line con 0
 exec-timeout 0 0
 logging synchronous
line aux 0
line vty 0 4
 login
!
!
end
```

4.1.3 Hub-and-Spoke IPSEC VPN – Static IP Spokes

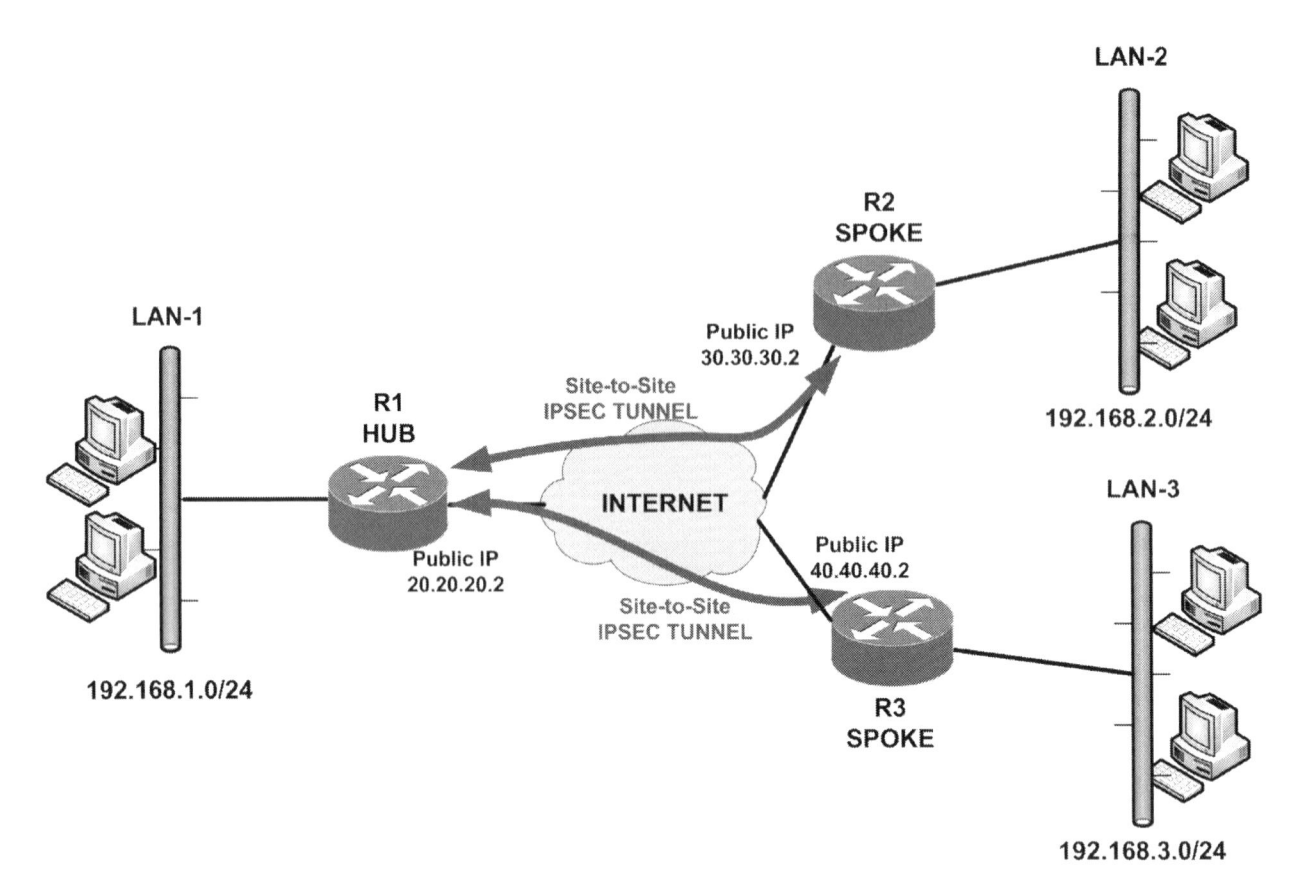

R1 (HUB)

version 12.4
service timestamps debug datetime msec
service timestamps log datetime msec
no service password-encryption
!
hostname R1
!
boot-start-marker
boot-end-marker
!
no aaa new-model
memory-size iomem 5
ip cef
!
!
ip auth-proxy max-nodata-conns 3
ip admission max-nodata-conns 3

!
!
crypto isakmp policy 1
 encr 3des
 hash md5
 authentication pre-share
crypto isakmp key secretkey1 address 30.30.30.2
crypto isakmp key secretkey2 address 40.40.40.2
!
!
crypto ipsec transform-set TRSET esp-3des esp-md5-hmac
!
crypto map VPNMAP 10 ipsec-isakmp
 set peer 30.30.30.2
 set transform-set TRSET
 match address VPN-TO-REMOTE1
crypto map VPNMAP 20 ipsec-isakmp
 set peer 40.40.40.2
 set transform-set TRSET
 match address VPN-TO-REMOTE2
!
!
interface FastEthernet0/0
 ip address 20.20.20.2 255.255.255.0
 ip nat outside
 ip virtual-reassembly
 speed 100
 full-duplex
 crypto map VPNMAP
!
interface FastEthernet0/1
 ip address 192.168.1.1 255.255.255.0
 ip nat inside
 ip virtual-reassembly
 duplex auto
 speed auto
!
ip forward-protocol nd
ip route 0.0.0.0 0.0.0.0 20.20.20.1
!
ip http server
no ip http secure-server
ip nat inside source list NAT-ACL interface FastEthernet0/0 overload
!
ip access-list extended NAT-ACL
 deny   ip 192.168.1.0 0.0.0.255 192.168.2.0 0.0.0.255

```
 deny   ip 192.168.1.0 0.0.0.255 192.168.3.0 0.0.0.255
 permit ip 192.168.1.0 0.0.0.255 any
ip access-list extended VPN-TO-REMOTE1
 permit ip 192.168.1.0 0.0.0.255 192.168.2.0 0.0.0.255
ip access-list extended VPN-TO-REMOTE2
 permit ip 192.168.1.0 0.0.0.255 192.168.3.0 0.0.0.255
!
control-plane
!
line con 0
line aux 0
line vty 0 4
 login
!
end
```

R2 (SPOKE)

```
version 12.4
service timestamps debug datetime msec
service timestamps log datetime msec
no service password-encryption
!
hostname R2
!
boot-start-marker
boot-end-marker
!
!
no aaa new-model
memory-size iomem 5
ip cef
!
!
ip auth-proxy max-nodata-conns 3
ip admission max-nodata-conns 3
!
 crypto isakmp policy 1
  encr 3des
  hash md5
  authentication pre-share
 crypto isakmp key secretkey1 address 20.20.20.2
 !
 !
```

```
crypto ipsec transform-set TRSET esp-3des esp-md5-hmac
!
crypto map VPNMAP 10 ipsec-isakmp
 set peer 20.20.20.2
 set transform-set TRSET
 match address VPN-TO-HQ
!
interface FastEthernet0/0
 ip address 30.30.30.2 255.255.255.0
 ip nat outside
 ip virtual-reassembly
 speed 100
 full-duplex
 crypto map VPNMAP
!
interface FastEthernet0/1
 ip address 192.168.2.1 255.255.255.0
 ip nat inside
 ip virtual-reassembly
 duplex auto
 speed auto
!
ip forward-protocol nd
ip route 0.0.0.0 0.0.0.0 30.30.30.1
!
ip http server
no ip http secure-server
ip nat inside source list NAT-ACL interface FastEthernet0/0 overload
!
ip access-list extended NAT-ACL
 deny   ip 192.168.2.0 0.0.0.255 192.168.1.0 0.0.0.255
 permit ip 192.168.2.0 0.0.0.255 any
ip access-list extended VPN-TO-HQ
 permit ip 192.168.2.0 0.0.0.255 192.168.1.0 0.0.0.255
!
control-plane
!
line con 0
line aux 0
line vty 0 4
 login
!
!
end
```

R3 (SPOKE)

```
version 12.4
service timestamps debug datetime msec
service timestamps log datetime msec
no service password-encryption
!
hostname R3
!
boot-start-marker
boot-end-marker
!
!
no aaa new-model
memory-size iomem 5
ip cef
!
!
ip auth-proxy max-nodata-conns 3
ip admission max-nodata-conns 3
!
!
no ip domain lookup
!
crypto isakmp policy 1
 encr 3des
 hash md5
 authentication pre-share
crypto isakmp key secretkey2 address 20.20.20.2
!
!
crypto ipsec transform-set TRSET esp-3des esp-md5-hmac
!
crypto map VPNMAP 10 ipsec-isakmp
 set peer 20.20.20.2
 set transform-set TRSET
 match address VPN-TO-HQ
!
!
!
interface FastEthernet0/0
 ip address 40.40.40.2 255.255.255.0
 ip nat outside
 ip virtual-reassembly
```

```
 speed 100
 full-duplex
 crypto map VPNMAP
!
interface FastEthernet0/1
 ip address 192.168.3.1 255.255.255.0
 ip nat inside
 ip virtual-reassembly
 duplex auto
 speed auto
!
ip forward-protocol nd
ip route 0.0.0.0 0.0.0.0 40.40.40.1
!
no ip http server
no ip http secure-server
ip nat inside source list NAT-ACL interface FastEthernet0/0 overload
!
ip access-list extended NAT-ACL
 deny   ip 192.168.3.0 0.0.0.255 192.168.1.0 0.0.0.255
 permit ip 192.168.3.0 0.0.0.255 any
ip access-list extended VPN-TO-HQ
 permit ip 192.168.3.0 0.0.0.255 192.168.1.0 0.0.0.255
!
control-plane
!
line con 0
 exec-timeout 0 0
 logging synchronous
line aux 0
line vty 0 4
 login
!
!
end
```

4.1.4 Hub-and-Spoke IPSEC VPN – Dynamic IP Spoke

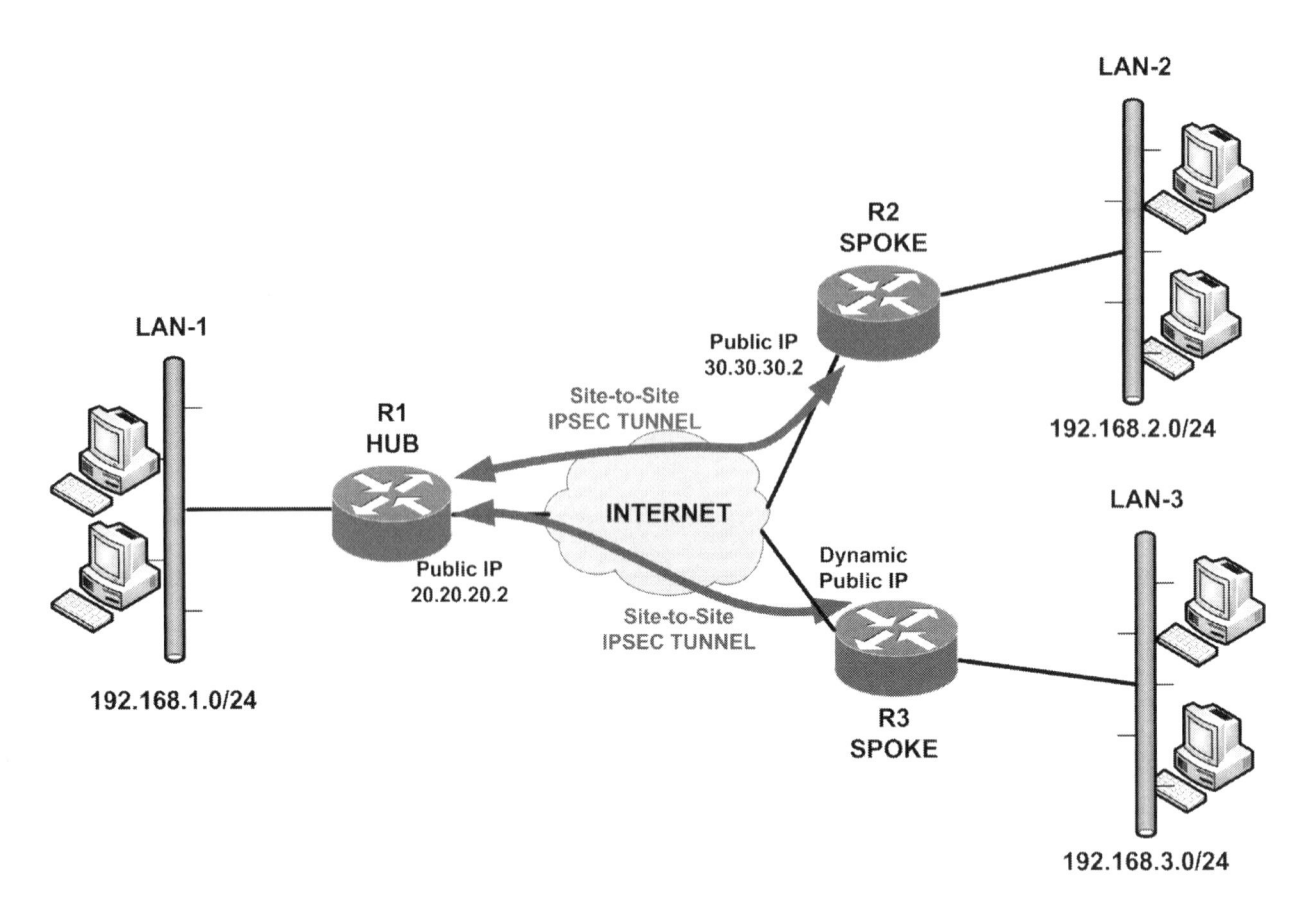

Only the Hub Configuration is shown since the Spokes are the same as the previous example.

<u>R1 (HUB)</u>

Current configuration : 1693 bytes
!
version 12.4
service timestamps debug datetime msec
service timestamps log datetime msec
no service password-encryption
!
hostname R1
!
boot-start-marker
boot-end-marker
!
!
no aaa new-model

```
 memory-size iomem 5
 ip cef
!
!
ip auth-proxy max-nodata-conns 3
ip admission max-nodata-conns 3
!
!
crypto isakmp policy 10
 encr 3des
 hash md5
 authentication pre-share
crypto isakmp key secretkey1 address 30.30.30.2
crypto isakmp key secretkey2 address 0.0.0.0 0.0.0.0
!
!
crypto ipsec transform-set TRSET esp-3des esp-md5-hmac
!
crypto dynamic-map dynmap 10
 set transform-set TRSET
 match address VPN-TO-REMOTE2
!
!
crypto map VPNMAP 10 ipsec-isakmp
 set peer 30.30.30.2
 set transform-set TRSET
 match address VPN-TO-REMOTE1
crypto map VPNMAP 20 ipsec-isakmp dynamic dynmap
!
!
!
interface FastEthernet0/0
 ip address 20.20.20.2 255.255.255.0
 ip nat outside
 ip virtual-reassembly
 crypto map VPNMAP
!
interface FastEthernet0/1
 ip address 192.168.1.1 255.255.255.0
 ip nat inside
 ip virtual-reassembly
 duplex auto
 speed auto
!
ip forward-protocol nd
ip route 0.0.0.0 0.0.0.0 20.20.20.1
```

```
!
ip http server
no ip http secure-server
ip nat inside source list NAT-ACL interface FastEthernet0/0 overload
!
ip access-list extended NAT-ACL
 deny   ip 192.168.1.0 0.0.0.255 192.168.2.0 0.0.0.255
 deny   ip 192.168.1.0 0.0.0.255 192.168.3.0 0.0.0.255
 permit ip 192.168.1.0 0.0.0.255 any
ip access-list extended VPN-TO-REMOTE1
 permit ip 192.168.1.0 0.0.0.255 192.168.2.0 0.0.0.255
ip access-list extended VPN-TO-REMOTE2
 permit ip 192.168.1.0 0.0.0.255 192.168.3.0 0.0.0.255
!
control-plane
!
line con 0
line aux 0
line vty 0 4
 login
!
!
end
```

4.1.5 Remote Access IPSEC VPN

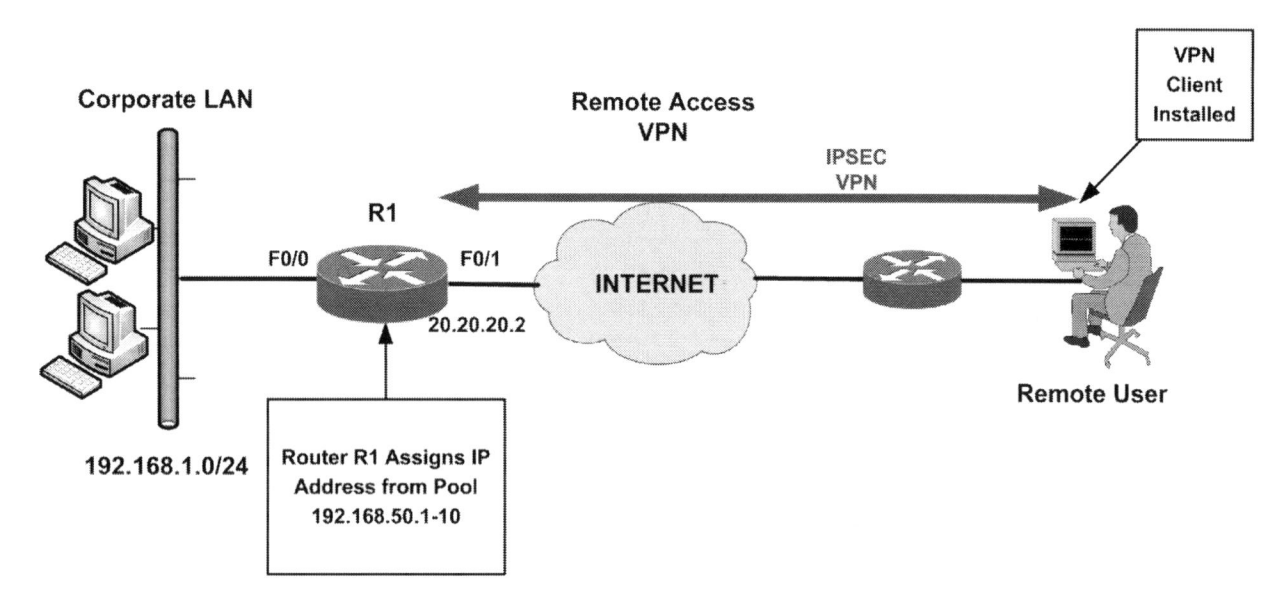

R1

version 12.4
service timestamps debug datetime msec
service timestamps log datetime msec
no service password-encryption
!
hostname R1
!
boot-start-marker
boot-end-marker
!
!
aaa new-model
!
!
aaa authentication login USERAUTH local
aaa authorization network NETAUTHORIZE local
!
aaa session-id common
memory-size iomem 5
ip cef
!
!
ip auth-proxy max-nodata-conns 3
ip admission max-nodata-conns 3

```
!
!
no ip domain lookup
!
username vpnuser password 0 cisco
!
crypto keyring vpnclientskey
  pre-shared-key address 0.0.0.0 0.0.0.0 key cisco123
!
crypto isakmp policy 10
 encr 3des
 authentication pre-share
 group 2
!
crypto isakmp client configuration group remotevpn
 key cisco123
 dns 192.168.1.2
 wins 192.168.1.2
 domain mycompany.com
 pool vpnpool
 acl VPN-ACL
crypto isakmp profile remoteclients
   description Remote Access VPN clients
   keyring vpnclientskey
   match identity group remotevpn
   client authentication list USERAUTH
   isakmp authorization list NETAUTHORIZE
   client configuration address respond
!
!
crypto ipsec transform-set TRSET esp-3des esp-md5-hmac
!
crypto dynamic-map dynmap 10
 set transform-set TRSET
 set isakmp-profile remoteclients
!
!
crypto map VPNMAP 10 ipsec-isakmp dynamic dynmap
!
interface FastEthernet0/0
 ip address 192.168.1.1 255.255.255.0
 ip nat inside
 ip virtual-reassembly
 duplex auto
 speed auto
!
```

```
interface FastEthernet0/1
 ip address 20.20.20.2 255.255.255.0
 ip nat outside
 ip virtual-reassembly
 speed 100
 full-duplex
 crypto map VPNMAP
!
ip local pool vpnpool 192.168.50.1 192.168.50.10
ip forward-protocol nd
ip route 0.0.0.0 0.0.0.0 20.20.20.1
!
no ip http server
no ip http secure-server
ip nat inside source list NAT-ACL interface FastEthernet0/1 overload
!
ip access-list extended NAT-ACL
 deny   ip 192.168.1.0 0.0.0.255 192.168.50.0 0.0.0.255
 permit ip 192.168.1.0 0.0.0.255 any
ip access-list extended VPN-ACL
 permit ip 192.168.1.0 0.0.0.255 192.168.50.0 0.0.0.255
!
control-plane
!
line con 0
 exec-timeout 0 0
 logging synchronous
line aux 0
line vty 0 4
!
!
end
```

4.1.6 Site-to-Site and Remote Access IPSEC VPN on same device

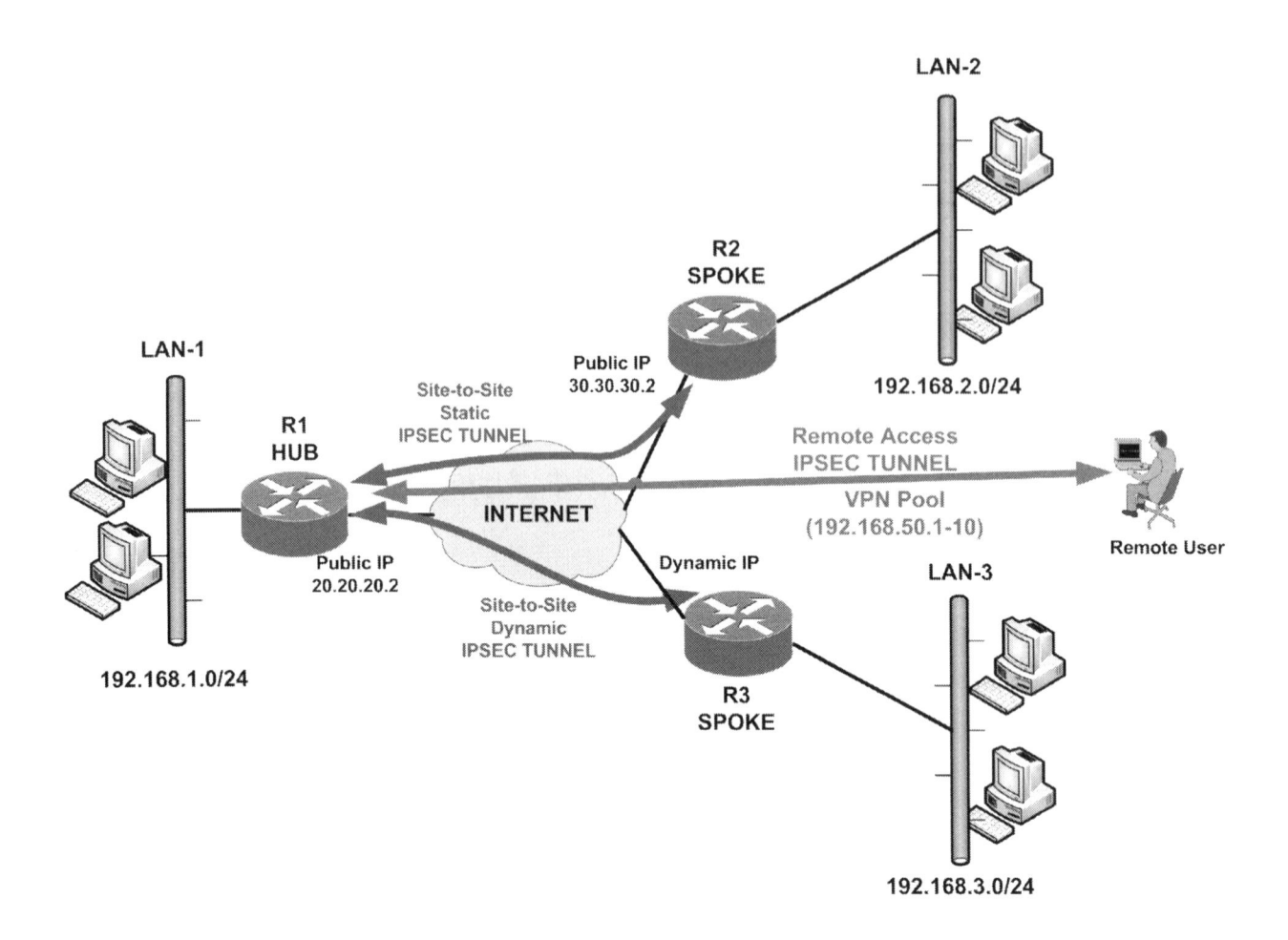

R1 (HUB)

version 12.4
service timestamps debug datetime msec
service timestamps log datetime msec
no service password-encryption
!
hostname R1
!
boot-start-marker
boot-end-marker
!
!
aaa new-model
!

```
!
aaa authentication login userauthen local
aaa authorization network groupauthor local
!
aaa session-id common
memory-size iomem 5
ip cef
!
ip auth-proxy max-nodata-conns 3
ip admission max-nodata-conns 3
!
!
no ip domain lookup
!
username vpnuser password 0 cisco
!
crypto keyring vpnclientskey
  pre-shared-key address 0.0.0.0 0.0.0.0 key cisco123
crypto keyring staticbranch
  pre-shared-key address 30.30.30.2 key secretkey1
crypto keyring dynamicbranch
  pre-shared-key address 0.0.0.0 0.0.0.0 key secretkey2
!
crypto isakmp policy 10
 encr 3des
 authentication pre-share
 group 2
!
crypto isakmp client configuration group remotevpn
 key cisco123
 dns 192.168.1.2
 wins 192.168.1.2
 domain mycompany.com
 pool vpnpool
 acl VPNclient-ACL
crypto isakmp profile remoteclients
   description Remote Access VPN clients
   keyring vpnclientskey
   match identity group remotevpn
   client authentication list userauthen
   isakmp authorization list groupauthor
   client configuration address respond
crypto isakmp profile staticL2L
   description isakmp profile for static Lan to Lan
   keyring staticbranch
   match identity address 30.30.30.2 255.255.255.255
```

```
crypto isakmp profile dynamicL2L
  description isakmp profile for dynamic Lan to Lan site
  keyring dynamicbranch
  match identity address 0.0.0.0
!
crypto ipsec transform-set TRSET esp-3des esp-md5-hmac
!
crypto dynamic-map DYNMAP 10
 set transform-set TRSET
 set isakmp-profile remoteclients
crypto dynamic-map DYNMAP 20
 set transform-set TRSET
 set isakmp-profile dynamicL2L
 match address VPNsite2-ACL
!
!
crypto map VPNMAP 10 ipsec-isakmp
 set peer 30.30.30.2
 set transform-set TRSET
 set isakmp-profile staticL2L
 match address VPNsite1-ACL
crypto map VPNMAP 20 ipsec-isakmp dynamic DYNMAP
!
interface FastEthernet0/0
 ip address 192.168.1.1 255.255.255.0
 ip nat inside
 ip virtual-reassembly
 duplex auto
 speed auto
!
interface FastEthernet0/1
 ip address 20.20.20.2 255.255.255.0
 ip nat outside
 ip virtual-reassembly
 speed 100
 full-duplex
 crypto map VPNMAP
!
ip local pool vpnpool 192.168.50.1 192.168.50.10
ip forward-protocol nd
ip route 0.0.0.0 0.0.0.0 20.20.20.1
!
no ip http server
no ip http secure-server
ip nat inside source list NAT-ACL interface FastEthernet0/1 overload
!
```

```
ip access-list extended NAT-ACL
 deny   ip 192.168.1.0 0.0.0.255 192.168.50.0 0.0.0.255
 deny   ip 192.168.1.0 0.0.0.255 192.168.2.0 0.0.0.255
 deny   ip 192.168.1.0 0.0.0.255 192.168.3.0 0.0.0.255
 permit ip 192.168.1.0 0.0.0.255 any

ip access-list extended VPNclient-ACL
 permit ip 192.168.1.0 0.0.0.255 192.168.50.0 0.0.0.255
ip access-list extended VPNsite1-ACL
 permit ip 192.168.1.0 0.0.0.255 192.168.2.0 0.0.0.255
ip access-list extended VPNsite2-ACL
 permit ip 192.168.1.0 0.0.0.255 192.168.3.0 0.0.0.255
!
control-plane
!
line con 0
line aux 0
line vty 0 4
!
!
end
```

R2 (SPOKE)

```
version 12.4
service timestamps debug datetime msec
service timestamps log datetime msec
no service password-encryption
!
hostname R2
!
boot-start-marker
boot-end-marker
!
!
no aaa new-model
memory-size iomem 5
ip cef
!
!
ip auth-proxy max-nodata-conns 3
ip admission max-nodata-conns 3
!
!
```

```
no ip domain lookup
!
!
crypto keyring hubsite
  pre-shared-key address 20.20.20.2 key secretkey1
!
crypto isakmp policy 10
 encr 3des
 authentication pre-share
 group 2
crypto isakmp profile staticL2L
   description isakmp profile for static Lan to Lan
   keyring hubsite
   match identity address 20.20.20.2 255.255.255.255
!
!
crypto ipsec transform-set TRSET esp-3des esp-md5-hmac
!
crypto map VPNMAP 5 ipsec-isakmp
 set peer 20.20.20.2
 set transform-set TRSET
 set isakmp-profile staticL2L
 match address VPN-ACL
!
!
!
interface FastEthernet0/0
 ip address 30.30.30.2 255.255.255.0
 ip nat outside
 ip virtual-reassembly
 speed 100
 full-duplex
 crypto map VPNMAP
!
interface FastEthernet0/1
 ip address 192.168.2.1 255.255.255.0
 ip nat inside
 ip virtual-reassembly
 speed 100
 full-duplex
!
ip forward-protocol nd
ip route 0.0.0.0 0.0.0.0 30.30.30.1
!
no ip http server
no ip http secure-server
```

```
ip nat inside source list NAT-ACL interface FastEthernet0/0 overload
!
ip access-list extended NAT-ACL
 deny   ip 192.168.2.0 0.0.0.255 192.168.1.0 0.0.0.255
 permit ip 192.168.2.0 0.0.0.255 any
ip access-list extended VPN-ACL
 permit ip 192.168.2.0 0.0.0.255 192.168.1.0 0.0.0.255
!
control-plane
!
!
line con 0
 exec-timeout 0 0
 logging synchronous
line aux 0
line vty 0 4
 login
!
!
end
```

R3 (SPOKE)

```
version 12.4
service timestamps debug datetime msec
service timestamps log datetime msec
no service password-encryption
!
hostname R3
!
boot-start-marker
boot-end-marker
!
!
no aaa new-model
memory-size iomem 5
ip cef
!
!
ip auth-proxy max-nodata-conns 3
ip admission max-nodata-conns 3
!
!
no ip domain lookup
!
```

```
crypto keyring hubsite
  pre-shared-key address 20.20.20.2 key secretkey2
!
crypto isakmp policy 10
 encr 3des
 authentication pre-share
 group 2
crypto isakmp profile dynamicL2L
   description isakmp profile for dynamic Lan to Lan site
   keyring hubsite
   match identity address 20.20.20.2 255.255.255.255
!
crypto ipsec transform-set TRSET esp-3des esp-md5-hmac
!
crypto map VPNMAP 5 ipsec-isakmp
 set peer 20.20.20.2
 set transform-set TRSET
 set isakmp-profile dynamicL2L
 match address VPN-ACL
!
interface FastEthernet0/0
 ip address dhcp
 ip nat outside
 ip virtual-reassembly
 speed 100
 full-duplex
 crypto map VPNMAP
!
interface FastEthernet0/1
 ip address 192.168.3.1 255.255.255.0
 ip nat inside
 ip virtual-reassembly
 duplex auto
 speed auto
!
ip forward-protocol nd
!
no ip http server
no ip http secure-server
ip nat inside source list NAT-ACL interface FastEthernet0/0 overload
!
ip access-list extended NAT-ACL
  deny   ip 192.168.3.0 0.0.0.255 192.168.1.0 0.0.0.255
  permit ip 192.168.3.0 0.0.0.255 any
 ip access-list extended VPN-ACL
  permit ip 192.168.3.0 0.0.0.255 192.168.1.0 0.0.0.255
```

```
!
control-plane
!
line con 0
 exec-timeout 0 0
 logging synchronous
line aux 0
line vty 0 4
 login
!
end
```

4.1.7 Site-to-Site VPN using GRE with IPSEC Protection

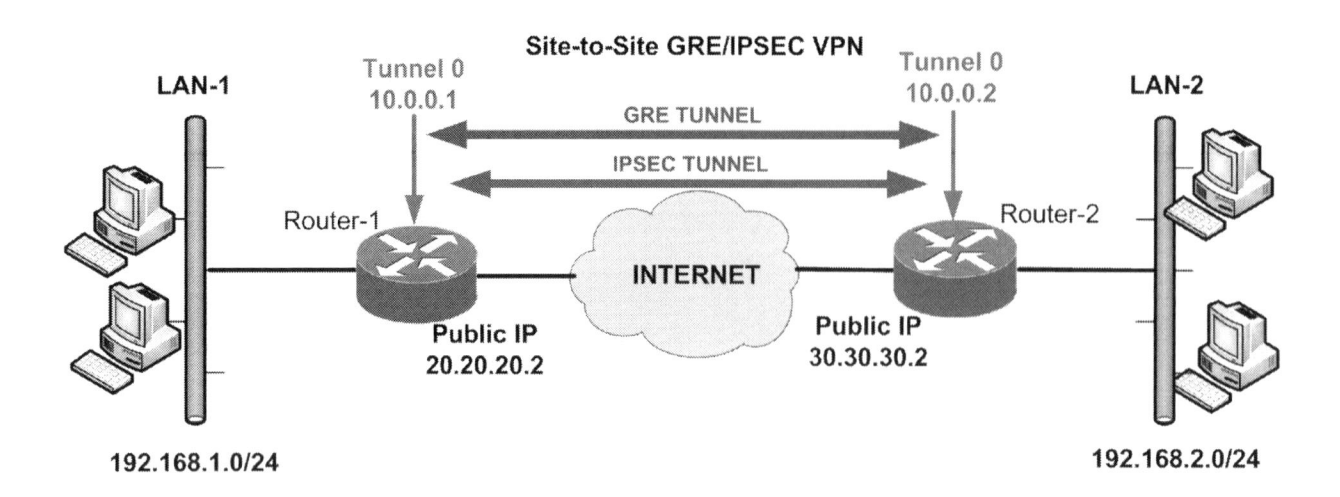

Router-1

version 12.4
service timestamps debug datetime msec
service timestamps log datetime msec
no service password-encryption
!
hostname R1
!
boot-start-marker
boot-end-marker
!
!
no aaa new-model
memory-size iomem 5
ip cef
!
!
ip auth-proxy max-nodata-conns 3
ip admission max-nodata-conns 3
!
no ip domain lookup
!
crypto isakmp policy 10
 encr 3des
 hash md5

```
 authentication pre-share
 group 2
crypto isakmp key testkey123 address 30.30.30.2
!
!
crypto ipsec transform-set TRSET esp-3des esp-md5-hmac
!
crypto ipsec profile GRE-PROTECTION
 set transform-set TRSET
!
!
interface Tunnel0
 ip address 10.0.0.1 255.255.255.0
 tunnel source 20.20.20.2
 tunnel destination 30.30.30.2
 tunnel protection ipsec profile GRE-PROTECTION
!
interface FastEthernet0/0
 ip address 20.20.20.2 255.255.255.0
 ip nat outside
 ip virtual-reassembly
 speed 100
 full-duplex
!
interface FastEthernet0/1
 ip address 192.168.1.1 255.255.255.0
 ip nat inside
 ip virtual-reassembly
 duplex auto
 speed auto
!
ip forward-protocol nd
ip route 0.0.0.0 0.0.0.0 20.20.20.1
ip route 192.168.2.0 255.255.255.0 10.0.0.2
!
no ip http server
no ip http secure-server
ip nat inside source list NAT-ACL interface FastEthernet0/0 overload
!
ip access-list extended NAT-ACL
 permit ip 192.168.1.0 0.0.0.255 any
!
control-plane
!
line con 0
 exec-timeout 0 0
```

```
 logging synchronous
line aux 0
line vty 0 4
 login
!
!
end
```

Router-2

```
version 12.4
service timestamps debug datetime msec
service timestamps log datetime msec
no service password-encryption
!
hostname R2
!
boot-start-marker
boot-end-marker
!
no aaa new-model
memory-size iomem 5
ip cef
!
!
ip auth-proxy max-nodata-conns 3
ip admission max-nodata-conns 3
!
!
no ip domain lookup
!
crypto isakmp policy 10
 encr 3des
 hash md5
 authentication pre-share
 group 2
crypto isakmp key testkey123 address 20.20.20.2
!
!
crypto ipsec transform-set TRSET esp-3des esp-md5-hmac
!
crypto ipsec profile GRE-PROTECTION
 set transform-set TRSET
!
```

```
interface Tunnel0
 ip address 10.0.0.2 255.255.255.0
 tunnel source 30.30.30.2
 tunnel destination 20.20.20.2
 tunnel protection ipsec profile GRE-PROTECTION
!
interface FastEthernet0/0
 ip address 30.30.30.2 255.255.255.0
 ip nat outside
 ip virtual-reassembly
 speed 100
 full-duplex
!
interface FastEthernet0/1
 ip address 192.168.2.1 255.255.255.0
 ip nat inside
 ip virtual-reassembly
 duplex auto
 speed auto
!
ip forward-protocol nd
ip route 0.0.0.0 0.0.0.0 30.30.30.1
ip route 192.168.1.0 255.255.255.0 10.0.0.1
!
no ip http server
no ip http secure-server
ip nat inside source list NAT-ACL interface FastEthernet0/0 overload
!
ip access-list extended NAT-ACL
 permit ip 192.168.2.0 0.0.0.255 any
!
!
!
control-plane
!
!
!
line con 0
 exec-timeout 0 0
 logging synchronous
line aux 0
line vty 0 4
 login
!
!
end
```

4.1.8 Hub-and-Spoke VPN using GRE with IPSEC Protection

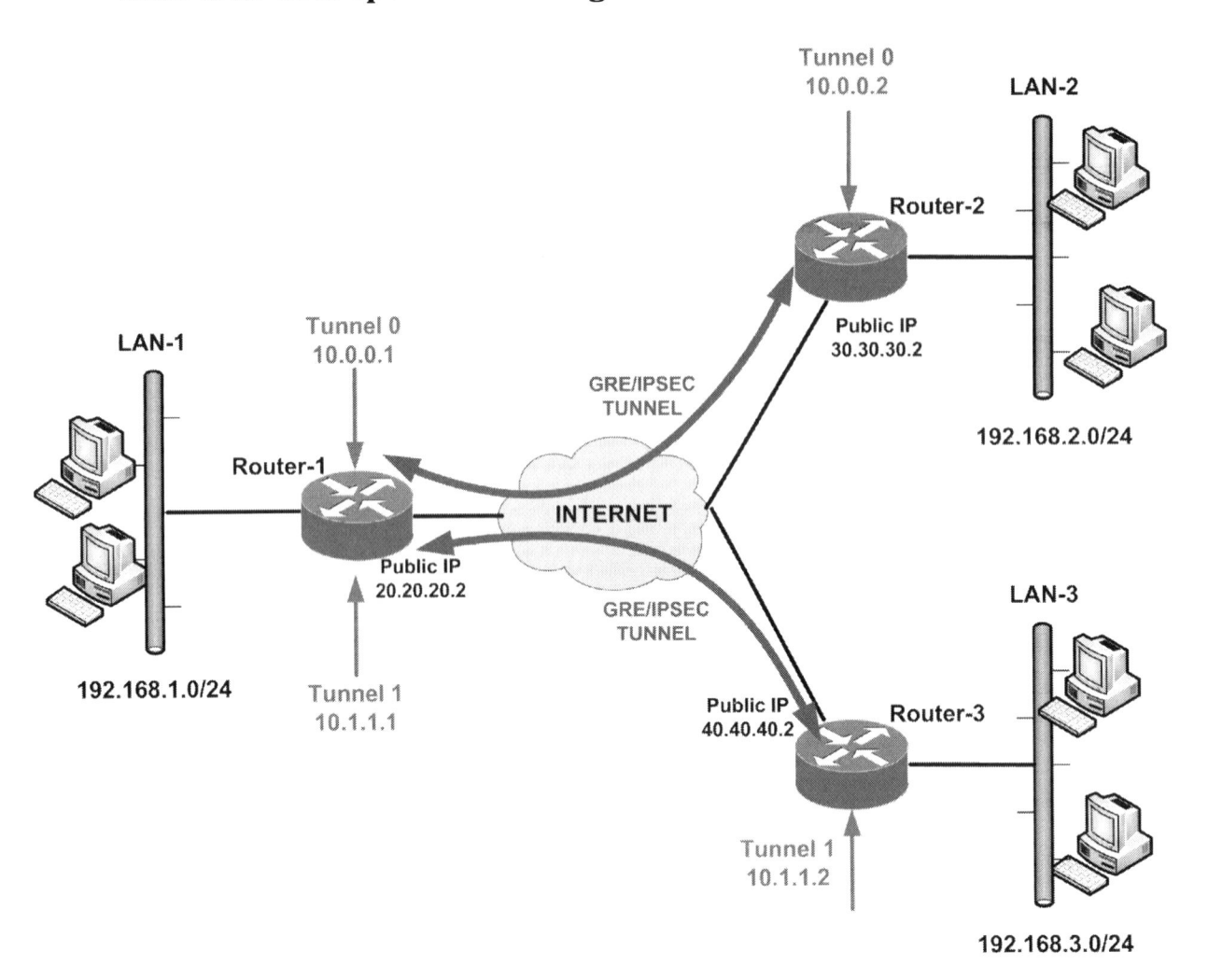

Router-1 (HUB)

version 12.4
service timestamps debug datetime msec
service timestamps log datetime msec
no service password-encryption
!
hostname R1
!
boot-start-marker
boot-end-marker
!
!
no aaa new-model
memory-size iomem 5

```
ip cef
!
!
ip auth-proxy max-nodata-conns 3
ip admission max-nodata-conns 3
!
no ip domain lookup
!
crypto isakmp policy 10
 encr 3des
 hash md5
 authentication pre-share
 group 2
crypto isakmp key testkey123 address 30.30.30.2
crypto isakmp key testkey1234 address 40.40.40.2
!
!
crypto ipsec transform-set TRSET esp-3des esp-md5-hmac
!
crypto ipsec profile GRE-PROTECTION
 set transform-set TRSET
!
!
interface Tunnel0
 ip address 10.0.0.1 255.255.255.0
 tunnel source 20.20.20.2
 tunnel destination 30.30.30.2
 tunnel protection ipsec profile GRE-PROTECTION
!
interface Tunnel1
 ip address 10.1.1.1 255.255.255.0
 tunnel source 20.20.20.2
 tunnel destination 40.40.40.2
 tunnel protection ipsec profile GRE-PROTECTION
!
interface FastEthernet0/0
 ip address 20.20.20.2 255.255.255.0
 ip nat outside
 ip virtual-reassembly
 speed 100
 full-duplex
!
interface FastEthernet0/1
 ip address 192.168.1.1 255.255.255.0
 ip nat inside
 ip virtual-reassembly
```

```
 duplex auto
 speed auto
!
!
router eigrp 100
 network 10.0.0.0 0.0.0.255
 network 10.1.1.0 0.0.0.255
 network 192.168.1.0
 no auto-summary
!
ip forward-protocol nd
ip route 0.0.0.0 0.0.0.0 20.20.20.1
!
no ip http server
no ip http secure-server
ip nat inside source list NAT-ACL interface FastEthernet0/0 overload
!
ip access-list extended NAT-ACL
 permit ip 192.168.1.0 0.0.0.255 any
!
control-plane
!
line con 0
 exec-timeout 0 0
 logging synchronous
line aux 0
line vty 0 4
 login
!
end
```

Router-2 (SPOKE)

```
version 12.4
service timestamps debug datetime msec
service timestamps log datetime msec
no service password-encryption
!
hostname R2
!
boot-start-marker
boot-end-marker
!
!
no aaa new-model
```

```
memory-size iomem 5
ip cef
!
!
ip auth-proxy max-nodata-conns 3
ip admission max-nodata-conns 3
!
!
no ip domain lookup
!
crypto isakmp policy 10
 encr 3des
 hash md5
 authentication pre-share
 group 2
crypto isakmp key testkey123 address 20.20.20.2
!
!
crypto ipsec transform-set TRSET esp-3des esp-md5-hmac
!
crypto ipsec profile GRE-PROTECTION
 set transform-set TRSET
!
!
interface Tunnel0
 ip address 10.0.0.2 255.255.255.0
 tunnel source 30.30.30.2
 tunnel destination 20.20.20.2
 tunnel protection ipsec profile GRE-PROTECTION
!
interface FastEthernet0/0
 ip address 30.30.30.2 255.255.255.0
 ip nat outside
 ip virtual-reassembly
 speed 100
 full-duplex
!
interface FastEthernet0/1
 ip address 192.168.2.1 255.255.255.0
 ip nat inside
 ip virtual-reassembly
 duplex auto
 speed auto
!
router eigrp 100
 network 10.0.0.0 0.0.0.255
```

```
 network 192.168.2.0
 no auto-summary
!
ip forward-protocol nd
ip route 0.0.0.0 0.0.0.0 30.30.30.1
!
no ip http server
no ip http secure-server
ip nat inside source list NAT-ACL interface FastEthernet0/0 overload
!
ip access-list extended NAT-ACL
 permit ip 192.168.2.0 0.0.0.255 any
control-plane
!
line con 0
 exec-timeout 0 0
 logging synchronous
line aux 0
line vty 0 4
 login
!
end
```

Router-3 (SPOKE)

```
version 12.4
service timestamps debug datetime msec
service timestamps log datetime msec
no service password-encryption
!
hostname R3
!
boot-start-marker
boot-end-marker
!
no aaa new-model
memory-size iomem 5
ip cef
!
ip auth-proxy max-nodata-conns 3
ip admission max-nodata-conns 3
!
!
 no ip domain lookup
```

```
!
!
crypto isakmp policy 10
 encr 3des
 hash md5
 authentication pre-share
 group 2
crypto isakmp key testkey1234 address 20.20.20.2
!
crypto ipsec transform-set TRSET esp-3des esp-md5-hmac
!
crypto ipsec profile GRE-PROTECTION
 set transform-set TRSET
interface Tunnel0
 ip address 10.1.1.2 255.255.255.0
 tunnel source 40.40.40.2
 tunnel destination 20.20.20.2
 tunnel protection ipsec profile GRE-PROTECTION
!
interface FastEthernet0/0
 ip address 40.40.40.2 255.255.255.0
 ip nat outside
 ip virtual-reassembly
 speed 100
 full-duplex
!
interface FastEthernet0/1
 ip address 192.168.3.1 255.255.255.0
 ip nat inside
 ip virtual-reassembly
 duplex auto
 speed auto
!
router eigrp 100
 network 10.1.1.0 0.0.0.255
 network 192.168.3.0
 no auto-summary
!
ip forward-protocol nd
ip route 0.0.0.0 0.0.0.0 40.40.40.1
!
no ip http server
no ip http secure-server
ip nat inside source list NAT-ACL interface FastEthernet0/0 overload
!
ip access-list extended NAT-ACL
```

```
 permit ip 192.168.3.0 0.0.0.255 any
!
control-plane
!
line con 0
 exec-timeout 0 0
 logging synchronous
line aux 0
line vty 0 4
 login
!
end
```

4.1.9 Hub-and-Spoke VPN using DVTI and SVTI

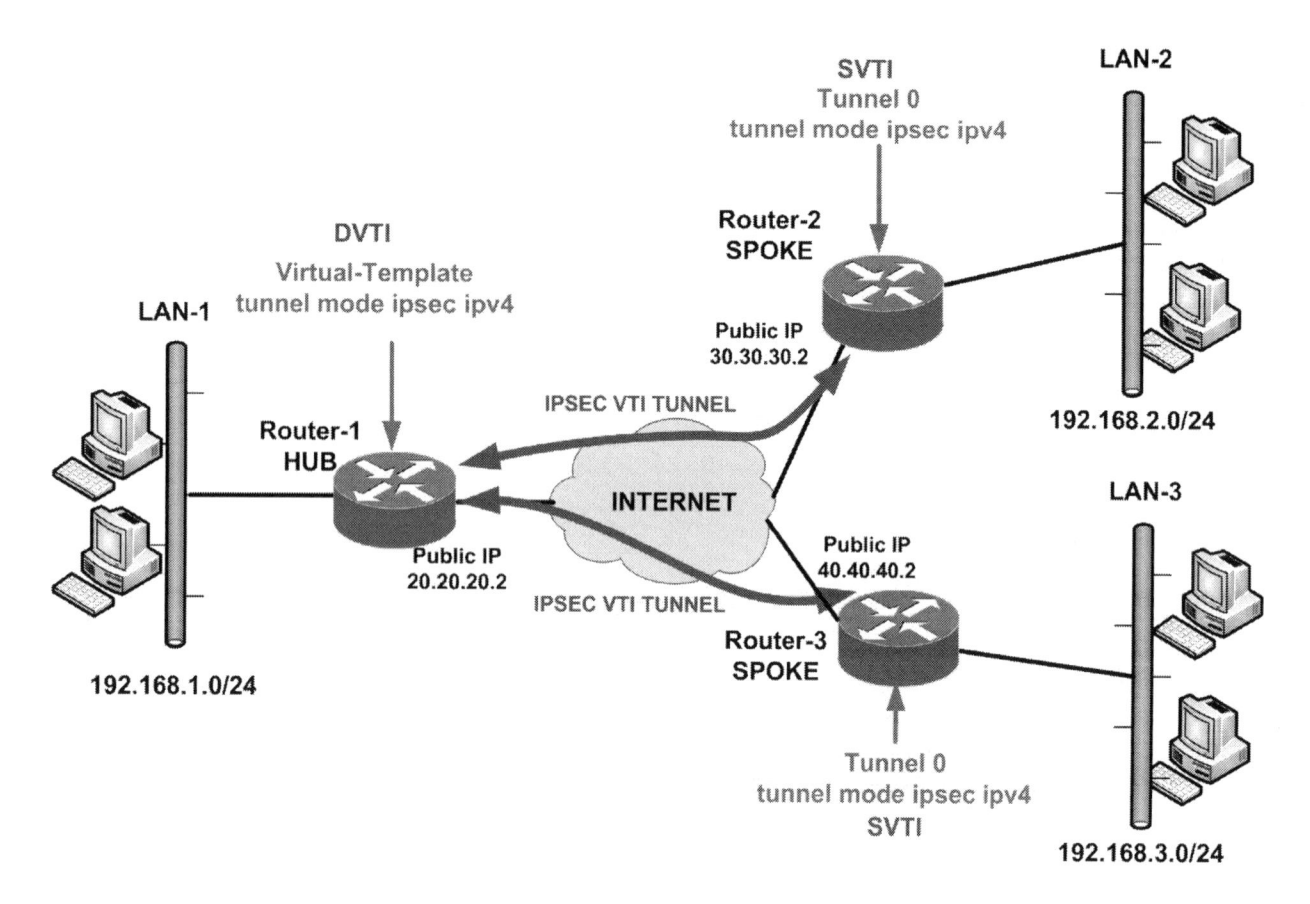

Router-1 (HUB)

```
version 12.4
service timestamps debug datetime msec
service timestamps log datetime msec
no service password-encryption
!
hostname R1
!
boot-start-marker
boot-end-marker
!
no logging console
!
no aaa new-model
memory-size iomem 5
ip cef
!
```

!
ip auth-proxy max-nodata-conns 3
ip admission max-nodata-conns 3
!
crypto keyring remotebranchkeys
  pre-shared-key address 0.0.0.0 0.0.0.0 key strongkey123
!
crypto isakmp policy 10
 encr 3des
 authentication pre-share
 group 2
crypto isakmp profile DVTI
   keyring remotebranchkeys
   match identity address 0.0.0.0
   virtual-template 1
!
!
crypto ipsec transform-set TRSET esp-3des esp-sha-hmac
!
crypto ipsec profile VTI-PROTECTION
 set transform-set TRSET
!
interface Loopback0
 ip address 10.0.0.1 255.255.255.0
!
interface FastEthernet0/0
 ip address 20.20.20.2 255.255.255.0
 ip nat outside
 ip virtual-reassembly
 speed 100
 full-duplex
!
interface FastEthernet0/1
 ip address 192.168.1.1 255.255.255.0
 ip nat inside
 ip virtual-reassembly
 duplex auto
 speed auto
!
 interface Virtual-Template1 type tunnel
  ip unnumbered Loopback0
  tunnel source FastEthernet0/0
  tunnel mode ipsec ipv4
  tunnel protection ipsec profile VTI-PROTECTION
  !
  router eigrp 100

```
 network 10.0.0.0 0.0.0.255
 network 192.168.1.0
 no auto-summary
!
ip forward-protocol nd
ip route 0.0.0.0 0.0.0.0 20.20.20.1
!
no ip http server
no ip http secure-server
ip nat inside source list NAT-ACL interface FastEthernet0/0 overload
!
ip access-list extended NAT-ACL
 permit ip 192.168.1.0 0.0.0.255 any
!
control-plane
!
line con 0
line aux 0
line vty 0 4
 login
!
!
end
```

Router-2 (SPOKE)

```
version 12.4
service timestamps debug datetime msec
service timestamps log datetime msec
no service password-encryption
!
hostname R2
!
boot-start-marker
boot-end-marker
!
!
no aaa new-model
memory-size iomem 5
ip cef
!
!
ip auth-proxy max-nodata-conns 3
ip admission max-nodata-conns 3
```

!
!
no ip domain lookup
!
!

crypto isakmp policy 10
 encr 3des
 authentication pre-share
 group 2

crypto isakmp key strongkey123 address 20.20.20.2
!
crypto ipsec transform-set TRSET esp-3des esp-sha-hmac
!
crypto ipsec profile VTI-PROTECTION
 set transform-set TRSET
!
interface Loopback0
 ip address 10.1.1.1 255.255.255.0
!
interface Tunnel0
 ip unnumbered Loopback0
 tunnel source FastEthernet0/0
 tunnel destination 20.20.20.2
 tunnel mode ipsec ipv4
 tunnel protection ipsec profile VTI-PROTECTION
!
interface FastEthernet0/0
 ip address 30.30.30.2 255.255.255.0
 ip nat outside
 ip virtual-reassembly
 speed 100
 full-duplex
!
interface FastEthernet0/1
 ip address 192.168.2.1 255.255.255.0
 ip nat inside
 ip virtual-reassembly
 speed 100
 full-duplex
!
router eigrp 100
 network 10.1.1.0 0.0.0.255
 network 192.168.2.0
 no auto-summary

!
ip forward-protocol nd
ip route 0.0.0.0 0.0.0.0 30.30.30.1
!
no ip http server
no ip http secure-server
ip nat inside source list NAT-ACL interface FastEthernet0/0 overload
!
ip access-list extended NAT-ACL
 permit ip 192.168.2.0 0.0.0.255 any
!
control-plane
!
line con 0
line aux 0
line vty 0 4
 login
!
!
end

Router-3 (SPOKE)

version 12.4
service timestamps debug datetime msec
service timestamps log datetime msec
no service password-encryption
!
hostname R3
!
boot-start-marker
boot-end-marker
!
no aaa new-model
memory-size iomem 5
ip cef
!
ip auth-proxy max-nodata-conns 3
ip admission max-nodata-conns 3
!
no ip domain lookup
!
crypto isakmp policy 10
 encr 3des
 authentication pre-share

group 2
crypto isakmp key strongkey123 address 20.20.20.2
!
crypto ipsec transform-set TRSET esp-3des esp-sha-hmac
!
crypto ipsec profile VTI-PROTECTION
 set transform-set TRSET
!
interface Loopback0
 ip address 10.2.2.1 255.255.255.0
interface Tunnel0
 ip unnumbered Loopback0
 tunnel source FastEthernet0/0
 tunnel destination 20.20.20.2
 tunnel mode ipsec ipv4
 tunnel protection ipsec profile VTI-PROTECTION
!
interface FastEthernet0/0
 ip address 40.40.40.2 255.255.255.0
 ip nat outside
 ip virtual-reassembly
 speed 100
 full-duplex
!
interface FastEthernet0/1
 ip address 192.168.3.1 255.255.255.0
 ip nat inside
 ip virtual-reassembly
 speed 100
 full-duplex
!
router eigrp 100
 network 10.2.2.0 0.0.0.255
 network 192.168.3.0
 no auto-summary
!
ip forward-protocol nd
ip route 0.0.0.0 0.0.0.0 40.40.40.1
!
no ip http server
no ip http secure-server
ip nat inside source list NAT-ACL interface FastEthernet0/0 overload
!
ip access-list extended NAT-ACL
 permit ip 192.168.3.0 0.0.0.255 any
!

```
control-plane
!
line con 0
 exec-timeout 0 0
 logging synchronous
line aux 0
line vty 0 4
 login
!
end
```

4.1.10 Dynamic Multipoint VPN (DMVPN)

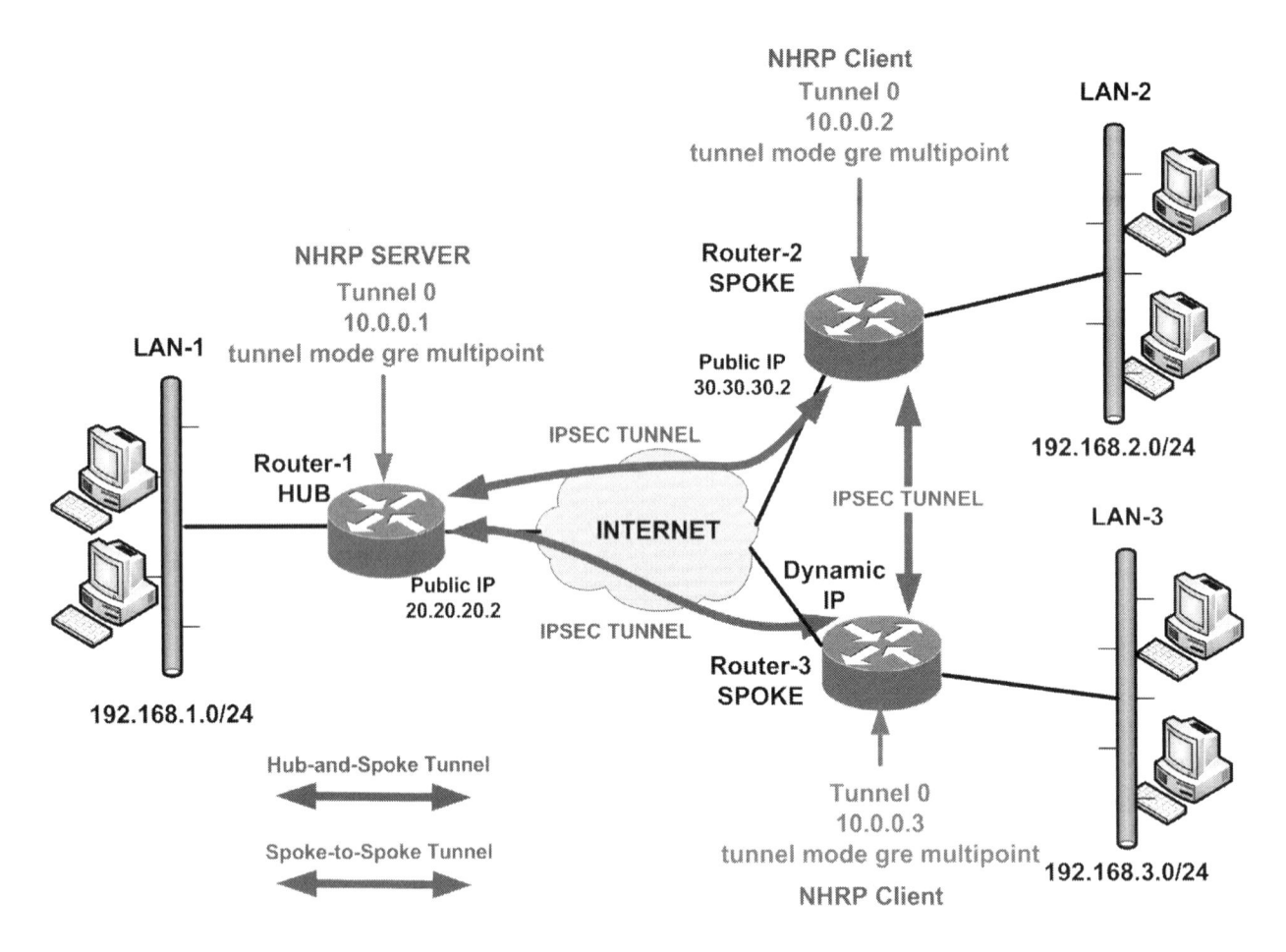

<u>Router-1 (HUB)</u>

version 12.4
service timestamps debug datetime msec
service timestamps log datetime msec
no service password-encryption
!
hostname R1
!
boot-start-marker
boot-end-marker
!
!
no aaa new-model
memory-size iomem 5
ip cef
!

!
ip auth-proxy max-nodata-conns 3
ip admission max-nodata-conns 3
!
!
no ip domain lookup
!
crypto isakmp policy 10
 encr 3des
 authentication pre-share
crypto isakmp key strongsecretkey address 0.0.0.0 0.0.0.0
crypto isakmp keepalive 10
!
crypto ipsec transform-set TRSET esp-3des esp-sha-hmac
!
crypto ipsec profile PROTECT-DMVPN
 set transform-set TRSET
!
interface Tunnel0
 ip address 10.0.0.1 255.255.255.0
 no ip redirects
 ip mtu 1440
 no ip next-hop-self eigrp 90
 ip nhrp authentication NHRPkey
 ip nhrp map multicast dynamic
 ip nhrp network-id 100
 no ip split-horizon eigrp 90
 tunnel source FastEthernet0/1
 tunnel mode gre multipoint
 tunnel protection ipsec profile PROTECT-DMVPN
!
interface FastEthernet0/0
 ip address 192.168.1.1 255.255.255.0
 ip nat inside
 ip virtual-reassembly
 speed 100
 full-duplex
!
interface FastEthernet0/1
 ip address 20.20.20.2 255.255.255.0
 ip nat outside
 ip virtual-reassembly
 speed 100
 full-duplex
!
router eigrp 90

```
 network 10.0.0.0 0.0.0.255
 network 192.168.1.0
 no auto-summary
!
ip forward-protocol nd
ip route 0.0.0.0 0.0.0.0 20.20.20.1
!
no ip http server
no ip http secure-server
ip nat inside source list NAT-ACL interface FastEthernet0/1 overload
!
ip access-list extended NAT-ACL
 permit ip 192.168.1.0 0.0.0.255 any
!
control-plane
!
line con 0
 exec-timeout 0 0
 logging synchronous
line aux 0
line vty 0 4
 login
!
!
end
```

Router-2 (SPOKE)

```
version 12.4
service timestamps debug datetime msec
service timestamps log datetime msec
no service password-encryption
!
hostname R2
!
boot-start-marker
boot-end-marker
!
!
no aaa new-model
memory-size iomem 5
ip cef
!
!
ip auth-proxy max-nodata-conns 3
```

```
ip admission max-nodata-conns 3
!
!
no ip domain lookup
!
!
crypto isakmp policy 10
 encr 3des
 authentication pre-share
crypto isakmp key strongsecretkey address 0.0.0.0 0.0.0.0
crypto isakmp keepalive 10
!
!
crypto ipsec transform-set TRSET esp-3des esp-sha-hmac
!
crypto ipsec profile PROTECT-DMVPN
 set transform-set TRSET
!
!
interface Tunnel0
 ip address 10.0.0.2 255.255.255.0
 no ip redirects
 ip mtu 1440
 no ip next-hop-self eigrp 90
 ip nhrp authentication NHRPkey
 ip nhrp map multicast dynamic
 ip nhrp map 10.0.0.1 20.20.20.2
 ip nhrp map multicast 20.20.20.2
 ip nhrp network-id 100
 ip nhrp nhs 10.0.0.1
 no ip split-horizon eigrp 90
 tunnel source FastEthernet0/1
 tunnel mode gre multipoint
 tunnel protection ipsec profile PROTECT-DMVPN
!
interface FastEthernet0/0
 ip address 192.168.2.1 255.255.255.0
 ip nat inside
 ip virtual-reassembly
 speed 100
 full-duplex
!
interface FastEthernet0/1
 ip address 30.30.30.2 255.255.255.0
 ip nat outside
 ip virtual-reassembly
```

```
 speed 100
 full-duplex
!
router eigrp 90
 network 10.0.0.0 0.0.0.255
 network 192.168.2.0
 no auto-summary
!
ip forward-protocol nd
ip route 0.0.0.0 0.0.0.0 30.30.30.1
!
no ip http server
no ip http secure-server
ip nat inside source list NAT-ACL interface FastEthernet0/1 overload
!
ip access-list extended NAT-ACL
 permit ip 192.168.2.0 0.0.0.255 any
!
control-plane
!
line con 0
 exec-timeout 0 0
 logging synchronous
line aux 0
line vty 0 4
 login
!
!
end
```

Router-3 (SPOKE)

```
version 12.4
service config
service timestamps debug datetime msec
service timestamps log datetime msec
no service password-encryption
!
hostname R3
!
boot-start-marker
boot-end-marker
!
no logging console
!
```

```
no aaa new-model
memory-size iomem 5
ip cef
!
!
ip auth-proxy max-nodata-conns 3
ip admission max-nodata-conns 3
!
!
no ip domain lookup
!
crypto isakmp policy 10
 encr 3des
 authentication pre-share
crypto isakmp key strongsecretkey address 0.0.0.0 0.0.0.0
crypto isakmp keepalive 10
!
!
crypto ipsec transform-set TRSET esp-3des esp-sha-hmac
!
crypto ipsec profile PROTECT-DMVPN
 set transform-set TRSET
!
interface Tunnel0
 ip address 10.0.0.3 255.255.255.0
 no ip redirects
 ip mtu 1440
 no ip next-hop-self eigrp 90
 ip nhrp authentication NHRPkey
 ip nhrp map multicast dynamic
 ip nhrp map 10.0.0.1 20.20.20.2
 ip nhrp map multicast 20.20.20.2
 ip nhrp network-id 100
 ip nhrp nhs 10.0.0.1
 no ip split-horizon eigrp 90
 tunnel source FastEthernet0/1
 tunnel mode gre multipoint
 tunnel protection ipsec profile PROTECT-DMVPN
!
interface FastEthernet0/0
 ip address 192.168.3.1 255.255.255.0
 ip nat inside
 ip virtual-reassembly
 speed 100
 full-duplex
!
```

```
interface FastEthernet0/1
 ip address dhcp
 ip nat outside
 ip virtual-reassembly
 speed 100
 full-duplex
!
router eigrp 90
 network 10.0.0.0 0.0.0.255
 network 192.168.3.0
 no auto-summary
!
ip forward-protocol nd
ip route 0.0.0.0 0.0.0.0 40.40.40.1
!
no ip http server
no ip http secure-server
ip nat inside source list NAT-ACL interface FastEthernet0/1 overload
!
ip access-list extended NAT-ACL
 permit ip 192.168.3.0 0.0.0.255 any
!
control-plane
!
line con 0
 exec-timeout 0 0
 logging synchronous
line aux 0
line vty 0 4
 login
!
!
end
```

4.1.11 Point to Point Tunelling Protocol (PPTP)

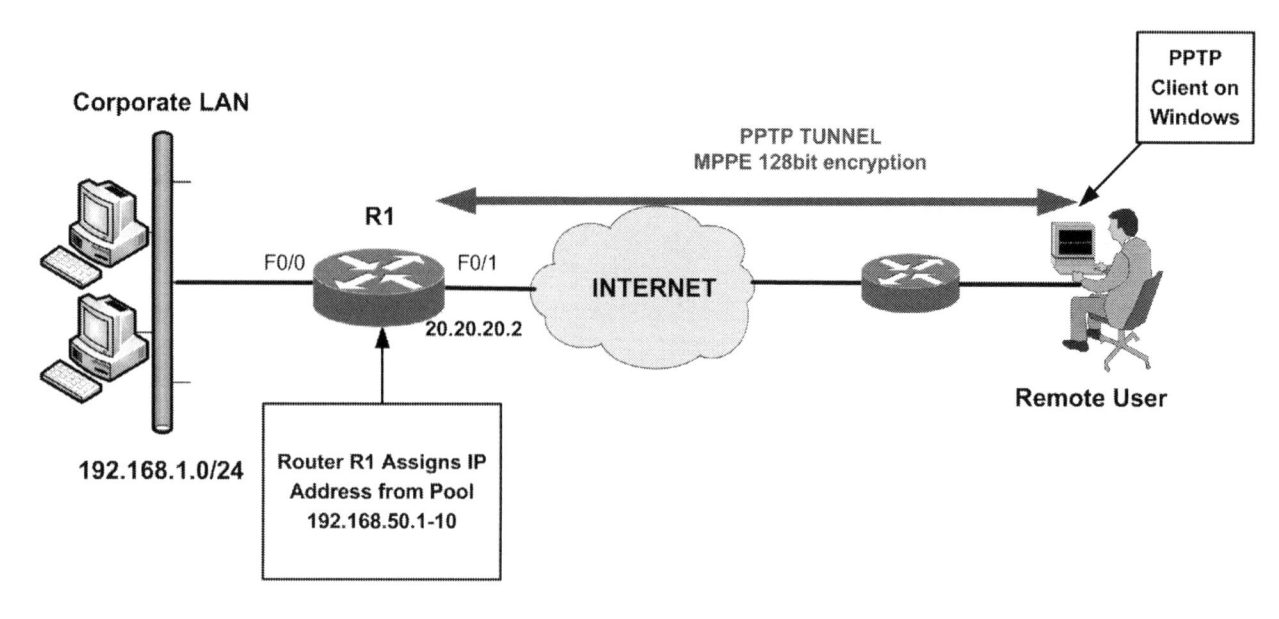

R1

```
version 12.4
service timestamps debug datetime msec
service timestamps log datetime msec
no service password-encryption
!
hostname R1
!
boot-start-marker
boot-end-marker
!
no aaa new-model
memory-size iomem 5
!
ip cef
no ip domain lookup
ip domain name lab.local
!
!
vpdn enable
!
vpdn-group 1
! Default PPTP VPDN group
 accept-dialin
  protocol pptp
```

```
 virtual-template 1
!
username remote1 password 0 cisco123
!
interface FastEthernet0/0
 ip address 192.168.1.1 255.255.255.0
 duplex auto
 speed auto
!
interface FastEthernet1/0
 ip address 20.20.20.2 255.255.255.0
 duplex auto
 speed auto
!
interface Virtual-Template1
 ip address 192.168.50.254 255.255.255.0
 peer default ip address pool pptp-pool
 ppp encrypt mppe 128
 ppp authentication ms-chap ms-chap-v2
!
ip local pool pptp-pool 192.168.50.1 192.168.50.10
no ip http server
no ip http secure-server
ip route 0.0.0.0 0.0.0.0 20.20.20.1
!
control-plane
!
line con 0
 exec-timeout 0 0
 privilege level 15
 logging synchronous
line aux 0
 exec-timeout 0 0
 privilege level 15
 logging synchronous
line vty 0 4
 login
!
!
end
```

4.2 Complete VPN Configurations on Cisco ASA

4.2.1 Site-to-Site IPSEC VPN

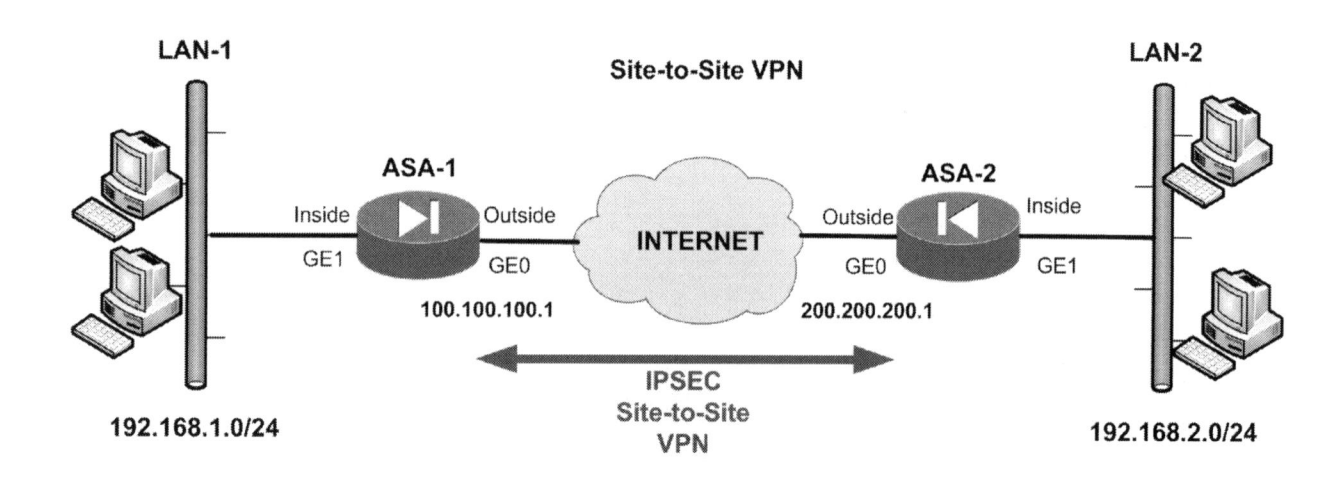

ASA-1

```
ASA Version 8.4(2)
!
hostname ASA1
enable password 8Ry2YjIyt7RRXU24 encrypted
passwd 2KFQnbNIdI.2KYOU encrypted
names
!
interface GigabitEthernet0
 nameif outside
 security-level 0
 ip address 100.100.100.1 255.255.255.0
!
interface GigabitEthernet1
 nameif inside
 security-level 100
 ip address 192.168.1.1 255.255.255.0
!
ftp mode passive
object network obj-local
 subnet 192.168.1.0 255.255.255.0
```

```
object network obj-remote
 subnet 192.168.2.0 255.255.255.0
object network internal-lan
 subnet 192.168.1.0 255.255.255.0
access-list outside_in extended permit icmp any any echo-reply
access-list outside_in extended deny ip any any log
access-list VPN-ACL extended permit ip 192.168.1.0 255.255.255.0 192.168.2.0 255.255.255.0
pager lines 24
mtu outside 1500
mtu inside 1500
icmp unreachable rate-limit 1 burst-size 1
no asdm history enable
arp timeout 14400
nat (inside,outside) source static obj-local obj-local destination static obj-remote obj-remote
!
object network internal-lan
 nat (inside,outside) dynamic interface
access-group outside_in in interface outside
route outside 0.0.0.0 0.0.0.0 100.100.100.2
timeout xlate 3:00:00
timeout conn 1:00:00 half-closed 0:10:00 udp 0:02:00 icmp 0:00:02
timeout sunrpc 0:10:00 h323 0:05:00 h225 1:00:00 mgcp 0:05:00 mgcp-pat 0:05:00
timeout sip 0:30:00 sip_media 0:02:00 sip-invite 0:03:00 sip-disconnect 0:02:00
timeout sip-provisional-media 0:02:00 uauth 0:05:00 absolute
timeout tcp-proxy-reassembly 0:01:00
timeout floating-conn 0:00:00
dynamic-access-policy-record DfltAccessPolicy
user-identity default-domain LOCAL
no snmp-server location
no snmp-server contact
snmp-server enable traps snmp authentication linkup linkdown coldstart warmstart
crypto ipsec ikev1 transform-set TRSET esp-aes esp-md5-hmac
crypto map VPNMAP 10 match address VPN-ACL
crypto map VPNMAP 10 set peer 200.200.200.1
crypto map VPNMAP 10 set ikev1 transform-set TRSET
crypto map VPNMAP interface outside
crypto isakmp identity address
crypto ikev1 enable outside
crypto ikev1 policy 10
 authentication pre-share
 encryption aes
 hash sha
 group 2
 lifetime 86400
telnet timeout 5
ssh timeout 5
```

console timeout 0
threat-detection basic-threat
threat-detection statistics access-list
no threat-detection statistics tcp-intercept
tunnel-group 200.200.200.1 type ipsec-l2l
tunnel-group 200.200.200.1 ipsec-attributes
 ikev1 pre-shared-key secretkey1
!
!
prompt hostname context
no call-home reporting anonymous
call-home
 profile CiscoTAC-1
  no active
  destination address http https://tools.cisco.com/its/service/oddce/services/DDCEService
  destination address email callhome@cisco.com
  destination transport-method http
  subscribe-to-alert-group diagnostic
  subscribe-to-alert-group environment
  subscribe-to-alert-group inventory periodic monthly
  subscribe-to-alert-group configuration periodic monthly
  subscribe-to-alert-group telemetry periodic daily
crashinfo save disable
Cryptochecksum:2e24c7d90262481b8fd7780418f9bfb6
: end

ASA-2

ASA Version 8.4(2)
!
hostname ASA2
enable password 8Ry2YjIyt7RRXU24 encrypted
passwd 2KFQnbNIdI.2KYOU encrypted
names
!
interface GigabitEthernet0
 nameif outside
 security-level 0
 ip address 200.200.200.1 255.255.255.0
!
interface GigabitEthernet1
 nameif inside

```
 security-level 100
 ip address 192.168.2.1 255.255.255.0
!
ftp mode passive
object network obj-local
 subnet 192.168.2.0 255.255.255.0
object network obj-remote
 subnet 192.168.1.0 255.255.255.0
object network internal-lan
 subnet 192.168.2.0 255.255.255.0
access-list outside_in extended permit icmp any any echo-reply
access-list outside_in extended deny ip any any log
access-list VPN-ACL extended permit ip 192.168.2.0 255.255.255.0 192.168.1.0 255.255.255.0
pager lines 24
mtu outside 1500
mtu inside 1500
icmp unreachable rate-limit 1 burst-size 1
no asdm history enable
arp timeout 14400
nat (inside,outside) source static obj-local obj-local destination static obj-remote obj-remote
!
object network internal-lan
 nat (inside,outside) dynamic interface
access-group outside_in in interface outside
route outside 0.0.0.0 0.0.0.0 200.200.200.2 1
timeout xlate 3:00:00
timeout conn 1:00:00 half-closed 0:10:00 udp 0:02:00 icmp 0:00:02
timeout sunrpc 0:10:00 h323 0:05:00 h225 1:00:00 mgcp 0:05:00 mgcp-pat 0:05:00
timeout sip 0:30:00 sip_media 0:02:00 sip-invite 0:03:00 sip-disconnect 0:02:00
timeout sip-provisional-media 0:02:00 uauth 0:05:00 absolute
timeout tcp-proxy-reassembly 0:01:00
timeout floating-conn 0:00:00
dynamic-access-policy-record DfltAccessPolicy
user-identity default-domain LOCAL
no snmp-server location
no snmp-server contact
snmp-server enable traps snmp authentication linkup linkdown coldstart warmstart
crypto ipsec ikev1 transform-set TRSET esp-aes esp-md5-hmac
crypto map VPNMAP 10 match address VPN-ACL
crypto map VPNMAP 10 set peer 100.100.100.1
crypto map VPNMAP 10 set ikev1 transform-set TRSET
crypto map VPNMAP interface outside
crypto isakmp identity address
crypto ikev1 enable outside
crypto ikev1 policy 10
 authentication pre-share
```

```
 encryption aes
 hash sha
 group 2
 lifetime 86400
telnet timeout 5
ssh timeout 5
console timeout 0
threat-detection basic-threat
threat-detection statistics access-list
no threat-detection statistics tcp-intercept
tunnel-group 100.100.100.1 type ipsec-l2l
tunnel-group 100.100.100.1 ipsec-attributes
 ikev1 pre-shared-key secretkey1
!
!
prompt hostname context
no call-home reporting anonymous
call-home
 profile CiscoTAC-1
  no active
  destination address http https://tools.cisco.com/its/service/oddce/services/DDCEService
  destination address email callhome@cisco.com
  destination transport-method http
  subscribe-to-alert-group diagnostic
  subscribe-to-alert-group environment
  subscribe-to-alert-group inventory periodic monthly
  subscribe-to-alert-group configuration periodic monthly
  subscribe-to-alert-group telemetry periodic daily
crashinfo save disable
Cryptochecksum:2e24c7d90262481b8fd7780418f9bfb6
: end
```

4.2.2 Hub-and-Spoke IPSEC VPN with Dynamic IP Spoke

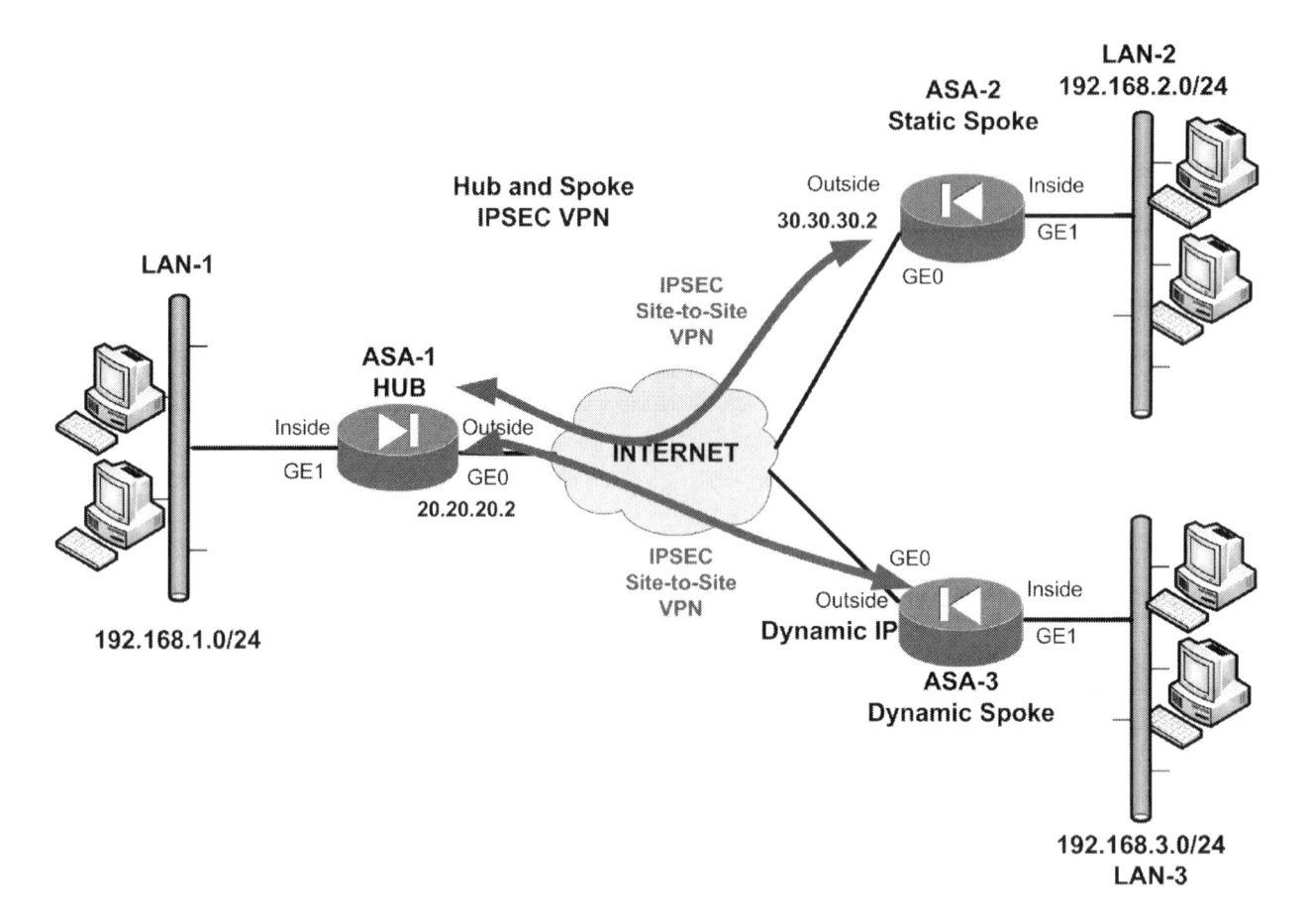

ASA-1 (HUB)

ASA Version 8.4(2)
!
hostname ASA1
enable password 8Ry2YjIyt7RRXU24 encrypted
passwd 2KFQnbNIdI.2KYOU encrypted
names
!
interface GigabitEthernet0
 nameif outside
 security-level 0
 ip address 20.20.20.2 255.255.255.0
!
interface GigabitEthernet1
 nameif inside
 security-level 100
 ip address 192.168.1.1 255.255.255.0

```
!
ftp mode passive
object network obj-local
 subnet 192.168.1.0 255.255.255.0
object network obj-remote1
 subnet 192.168.2.0 255.255.255.0
object network obj-remote2
 subnet 192.168.3.0 255.255.255.0
object network internal-lan
 subnet 192.168.1.0 255.255.255.0
access-list outside_in extended permit icmp any any echo-reply
access-list outside_in extended deny ip any any log
access-list VPN-ACL1 extended permit ip 192.168.1.0 255.255.255.0 192.168.2.0 255.255.255.0
access-list VPN-ACL2 extended permit ip 192.168.1.0 255.255.255.0 192.168.3.0 255.255.255.0
pager lines 24
mtu outside 1500
mtu inside 1500
icmp unreachable rate-limit 1 burst-size 1
no asdm history enable
arp timeout 14400
nat (inside,outside) 1 source static obj-local obj-local destination static obj-remote1 obj-remote1
nat (inside,outside) 2 source static obj-local obj-local destination static obj-remote2 obj-remote2
!
object network internal-lan
 nat (inside,outside) dynamic interface
access-group outside_in in interface outside
route outside 0.0.0.0 0.0.0.0 20.20.20.1 1
timeout xlate 3:00:00
timeout conn 1:00:00 half-closed 0:10:00 udp 0:02:00 icmp 0:00:02
timeout sunrpc 0:10:00 h323 0:05:00 h225 1:00:00 mgcp 0:05:00 mgcp-pat 0:05:00
timeout sip 0:30:00 sip_media 0:02:00 sip-invite 0:03:00 sip-disconnect 0:02:00
timeout sip-provisional-media 0:02:00 uauth 0:05:00 absolute
timeout tcp-proxy-reassembly 0:01:00
timeout floating-conn 0:00:00
dynamic-access-policy-record DfltAccessPolicy
user-identity default-domain LOCAL
no snmp-server location
no snmp-server contact
snmp-server enable traps snmp authentication linkup linkdown coldstart warmstart
crypto ipsec ikev1 transform-set TRSET esp-3des esp-md5-hmac
crypto dynamic-map DYNMAP 10 match address VPN-ACL2
crypto dynamic-map DYNMAP 10 set ikev1 transform-set TRSET
crypto map VPNMAP 5 match address VPN-ACL1
crypto map VPNMAP 5 set peer 30.30.30.2
crypto map VPNMAP 5 set ikev1 transform-set TRSET
crypto map VPNMAP 10 ipsec-isakmp dynamic DYNMAP
```

```
crypto map VPNMAP interface outside
crypto isakmp identity address
crypto ikev1 enable outside
crypto ikev1 policy 10
 authentication pre-share
 encryption 3des
 hash sha
 group 2
 lifetime 86400
telnet timeout 5
ssh timeout 5
console timeout 0
threat-detection basic-threat
threat-detection statistics access-list
no threat-detection statistics tcp-intercept
tunnel-group DefaultL2LGroup ipsec-attributes
 ikev1 pre-shared-key secretkey2
tunnel-group 30.30.30.2 type ipsec-l2l
tunnel-group 30.30.30.2 ipsec-attributes
 ikev1 pre-shared-key secretkey1
!
!
prompt hostname context
no call-home reporting anonymous
call-home
 profile CiscoTAC-1
  no active
  destination address http https://tools.cisco.com/its/service/oddce/services/DDCEService
  destination address email callhome@cisco.com
  destination transport-method http
  subscribe-to-alert-group diagnostic
  subscribe-to-alert-group environment
  subscribe-to-alert-group inventory periodic monthly
  subscribe-to-alert-group configuration periodic monthly
  subscribe-to-alert-group telemetry periodic daily
crashinfo save disable
Cryptochecksum:d29a9a3dc0e56bf1c5a174e88a4afc72
: end
```

ASA-2 (Static IP Spoke)

```
ASA Version 8.4(2)
!
hostname ASA2
enable password 8Ry2YjIyt7RRXU24 encrypted
passwd 2KFQnbNIdI.2KYOU encrypted
names
!
interface GigabitEthernet0
 nameif outside
 security-level 0
 ip address 30.30.30.2 255.255.255.0
!
interface GigabitEthernet1
 nameif inside
 security-level 100
 ip address 192.168.2.1 255.255.255.0
!
ftp mode passive
object network obj-local
 subnet 192.168.2.0 255.255.255.0
object network obj-remote
 subnet 192.168.1.0 255.255.255.0
object network internal-lan
 subnet 192.168.2.0 255.255.255.0
access-list outside_in extended permit icmp any any echo-reply
access-list outside_in extended deny ip any any log
access-list VPN-ACL extended permit ip 192.168.2.0 255.255.255.0 192.168.1.0 255.255.255.0
pager lines 24
mtu outside 1500
mtu inside 1500
icmp unreachable rate-limit 1 burst-size 1
no asdm history enable
arp timeout 14400
nat (inside,outside) source static obj-local obj-local destination static obj-remote obj-remote
!
object network internal-lan
 nat (inside,outside) dynamic interface
access-group outside_in in interface outside
route outside 0.0.0.0 0.0.0.0 30.30.30.1 1
timeout xlate 3:00:00
timeout conn 1:00:00 half-closed 0:10:00 udp 0:02:00 icmp 0:00:02
timeout sunrpc 0:10:00 h323 0:05:00 h225 1:00:00 mgcp 0:05:00 mgcp-pat 0:05:00
timeout sip 0:30:00 sip_media 0:02:00 sip-invite 0:03:00 sip-disconnect 0:02:00
timeout sip-provisional-media 0:02:00 uauth 0:05:00 absolute
```

```
timeout tcp-proxy-reassembly 0:01:00
timeout floating-conn 0:00:00
dynamic-access-policy-record DfltAccessPolicy
user-identity default-domain LOCAL
no snmp-server location
no snmp-server contact
snmp-server enable traps snmp authentication linkup linkdown coldstart warmstart
crypto ipsec ikev1 transform-set TRSET esp-3des esp-md5-hmac
crypto map VPNMAP 5 match address VPN-ACL
crypto map VPNMAP 5 set peer 20.20.20.2
crypto map VPNMAP 5 set ikev1 transform-set TRSET
crypto map VPNMAP interface outside
crypto isakmp identity address
crypto ikev1 enable outside
crypto ikev1 policy 10
 authentication pre-share
 encryption 3des
 hash sha
 group 2
 lifetime 86400
telnet timeout 5
ssh timeout 5
console timeout 0
threat-detection basic-threat
threat-detection statistics access-list
no threat-detection statistics tcp-intercept
tunnel-group 20.20.20.2 type ipsec-l2l
tunnel-group 20.20.20.2 ipsec-attributes
 ikev1 pre-shared-key secretkey1
!
!
prompt hostname context
no call-home reporting anonymous
call-home
 profile CiscoTAC-1
  no active
  destination address http https://tools.cisco.com/its/service/oddce/services/DDCEService
  destination address email callhome@cisco.com
  destination transport-method http
  subscribe-to-alert-group diagnostic
  subscribe-to-alert-group environment
  subscribe-to-alert-group inventory periodic monthly
  subscribe-to-alert-group configuration periodic monthly
  subscribe-to-alert-group telemetry periodic daily
 crashinfo save disable
Cryptochecksum:2e24c7d90262481b8fd7780418f9bfb6
```

ASA-3 (Dynamic IP Spoke)

```
ASA Version 8.4(2)
!
hostname ASA3
enable password 8Ry2Yjlyt7RRXU24 encrypted
passwd 2KFQnbNIdI.2KYOU encrypted
names
!
interface GigabitEthernet0
 nameif outside
 security-level 0
 ip address dhcp setroute
!
interface GigabitEthernet1
 nameif inside
 security-level 100
 ip address 192.168.3.1 255.255.255.0
!
ftp mode passive
object network obj-local
 subnet 192.168.3.0 255.255.255.0
object network obj-remote
 subnet 192.168.1.0 255.255.255.0
object network internal-lan
 subnet 192.168.3.0 255.255.255.0
access-list outside_in extended permit icmp any any echo-reply
access-list outside_in extended deny ip any any log
access-list VPN-ACL extended permit ip 192.168.3.0 255.255.255.0 192.168.1.0 255.255.255.0
pager lines 24
mtu outside 1500
mtu inside 1500
icmp unreachable rate-limit 1 burst-size 1
no asdm history enable
arp timeout 14400
nat (inside,outside) source static obj-local obj-local destination static obj-remote obj-remote
!
object network internal-lan
 nat (inside,outside) dynamic interface
access-group outside_in in interface outside
timeout xlate 3:00:00
timeout conn 1:00:00 half-closed 0:10:00 udp 0:02:00 icmp 0:00:02
timeout sunrpc 0:10:00 h323 0:05:00 h225 1:00:00 mgcp 0:05:00 mgcp-pat 0:05:00
timeout sip 0:30:00 sip_media 0:02:00 sip-invite 0:03:00 sip-disconnect 0:02:00
timeout sip-provisional-media 0:02:00 uauth 0:05:00 absolute
```

```
timeout tcp-proxy-reassembly 0:01:00
timeout floating-conn 0:00:00
dynamic-access-policy-record DfltAccessPolicy
user-identity default-domain LOCAL
no snmp-server location
no snmp-server contact
snmp-server enable traps snmp authentication linkup linkdown coldstart warmstart
crypto ipsec ikev1 transform-set TRSET esp-3des esp-md5-hmac
crypto map VPNMAP 5 match address VPN-ACL
crypto map VPNMAP 5 set peer 20.20.20.2
crypto map VPNMAP 5 set ikev1 transform-set TRSET
crypto map VPNMAP interface outside
crypto isakmp identity address
crypto ikev1 enable outside
crypto ikev1 policy 10
 authentication pre-share
 encryption 3des
 hash sha
 group 2
 lifetime 86400
telnet timeout 5
ssh timeout 5
console timeout 0
threat-detection basic-threat
threat-detection statistics access-list
no threat-detection statistics tcp-intercept
tunnel-group 20.20.20.2 type ipsec-l2l
tunnel-group 20.20.20.2 ipsec-attributes
 ikev1 pre-shared-key secretkey2
!
!
prompt hostname context
no call-home reporting anonymous
call-home
 profile CiscoTAC-1
  no active
  destination address http https://tools.cisco.com/its/service/oddce/services/DDCEService
  destination address email callhome@cisco.com
  destination transport-method http
  subscribe-to-alert-group diagnostic
  subscribe-to-alert-group environment
  subscribe-to-alert-group inventory periodic monthly
  subscribe-to-alert-group configuration periodic monthly
  subscribe-to-alert-group telemetry periodic daily
crashinfo save disable
Cryptochecksum:2e24c7d90262481b8fd7780418f9bfb6
```

4.2.3 IPSEC VPN Between Cisco ASA and Cisco Router

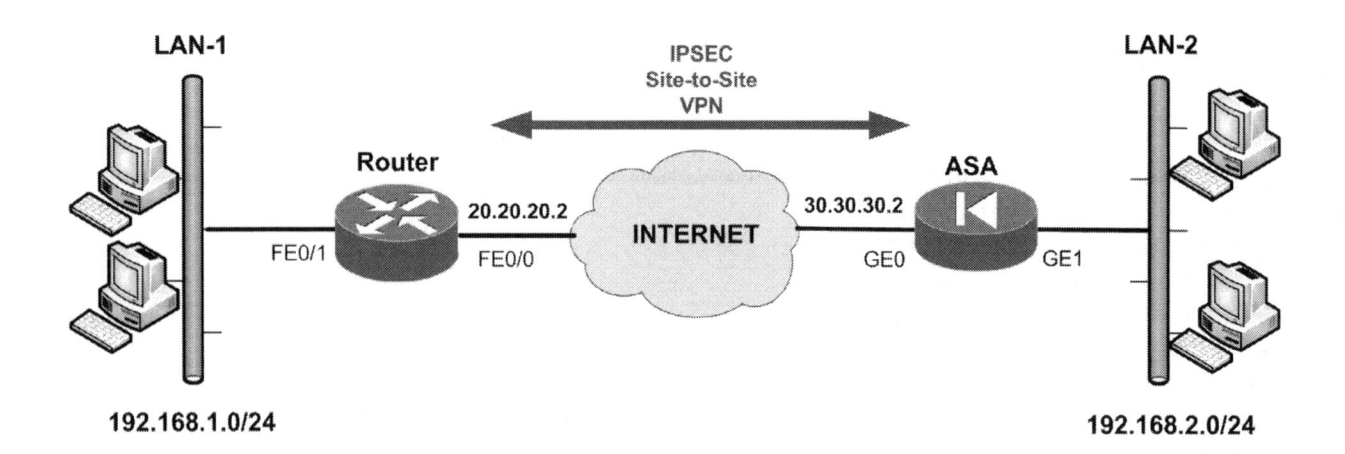

ROUTER

```
version 12.4
service timestamps debug datetime msec
service timestamps log datetime msec
no service password-encryption
!
hostname R1
!
boot-start-marker
boot-end-marker
!
no aaa new-model
memory-size iomem 5
ip cef
!
ip auth-proxy max-nodata-conns 3
ip admission max-nodata-conns 3
!
no ip domain lookup
!
crypto keyring ASAVPNKEY
  pre-shared-key address 30.30.30.2 key secretkey1
!
crypto isakmp policy 10
 encr 3des
 hash md5
 authentication pre-share
 group 2
```

!
crypto isakmp profile staticL2L
   description isakmp profile for static Lan to Lan with ASA
   keyring ASAVPNKEY
   match identity address 30.30.30.2 255.255.255.255
!
crypto ipsec transform-set TRSET esp-3des esp-md5-hmac
!
crypto map VPNMAP 10 ipsec-isakmp
 set peer 30.30.30.2
 set transform-set TRSET
 set isakmp-profile staticL2L
 match address VPN-ACL
!
interface FastEthernet0/0
 ip address 20.20.20.2 255.255.255.0
 ip nat outside
 ip virtual-reassembly
 crypto map VPNMAP
!
interface FastEthernet0/1
 ip address 192.168.1.1 255.255.255.0
 ip nat inside
 ip virtual-reassembly
 duplex auto
 speed auto
!
ip forward-protocol nd
ip route 0.0.0.0 0.0.0.0 20.20.20.1
!
no ip http server
no ip http secure-server
ip nat inside source list NAT-ACL interface FastEthernet0/0 overload
!
ip access-list extended NAT-ACL
 deny   ip 192.168.1.0 0.0.0.255 192.168.2.0 0.0.0.255
 permit ip 192.168.1.0 0.0.0.255 any
ip access-list extended VPN-ACL
 permit ip 192.168.1.0 0.0.0.255 192.168.2.0 0.0.0.255
!
 control-plane
!
 line con 0
  exec-timeout 0 0
  logging synchronous

```
 line aux 0
 line vty 0 4
  login
```

ASA

```
ASA Version 8.4(2)
!
hostname ASA
enable password 8Ry2YjIyt7RRXU24 encrypted
passwd 2KFQnbNIdI.2KYOU encrypted
names
!
interface GigabitEthernet0
 nameif outside
 security-level 0
 ip address 30.30.30.2 255.255.255.0
!
interface GigabitEthernet1
 nameif inside
 security-level 100
 ip address 192.168.2.1 255.255.255.0
!
ftp mode passive
object network obj-local
 subnet 192.168.2.0 255.255.255.0
object network obj-remote
 subnet 192.168.1.0 255.255.255.0
object network internal-lan
 subnet 192.168.2.0 255.255.255.0
access-list outside_in extended permit icmp any any echo-reply
access-list outside_in extended deny ip any any log
access-list VPN-ACL extended permit ip 192.168.2.0 255.255.255.0 192.168.1.0 255.255.255.0
pager lines 24
mtu outside 1500
mtu inside 1500
icmp unreachable rate-limit 1 burst-size 1
no asdm history enable
arp timeout 14400
nat (inside,outside) source static obj-local obj-local destination static obj-remote obj-remote
!
object network internal-lan
 nat (inside,outside) dynamic interface

access-group outside_in in interface outside
route outside 0.0.0.0 0.0.0.0 30.30.30.1 1
```

timeout xlate 3:00:00
timeout conn 1:00:00 half-closed 0:10:00 udp 0:02:00 icmp 0:00:02
timeout sunrpc 0:10:00 h323 0:05:00 h225 1:00:00 mgcp 0:05:00 mgcp-pat 0:05:00
timeout sip 0:30:00 sip_media 0:02:00 sip-invite 0:03:00 sip-disconnect 0:02:00
timeout sip-provisional-media 0:02:00 uauth 0:05:00 absolute
timeout tcp-proxy-reassembly 0:01:00
timeout floating-conn 0:00:00
dynamic-access-policy-record DfltAccessPolicy
user-identity default-domain LOCAL
no snmp-server location
no snmp-server contact
snmp-server enable traps snmp authentication linkup linkdown coldstart warmstart
crypto ipsec ikev1 transform-set TRSET esp-3des esp-md5-hmac
crypto map VPNMAP 10 match address VPN-ACL
crypto map VPNMAP 10 set peer 20.20.20.2
crypto map VPNMAP 10 set ikev1 transform-set TRSET
crypto map VPNMAP interface outside
crypto isakmp identity address
crypto ikev1 enable outside
crypto ikev1 policy 10
 authentication pre-share
 encryption 3des
 hash md5
 group 2
 lifetime 86400
telnet timeout 5
ssh timeout 5
console timeout 0
threat-detection basic-threat
threat-detection statistics access-list
no threat-detection statistics tcp-intercept
tunnel-group 20.20.20.2 type ipsec-l2l
tunnel-group 20.20.20.2 ipsec-attributes
 ikev1 pre-shared-key secretkey1
!
!
prompt hostname context
no call-home reporting anonymous
call-home
 profile CiscoTAC-1
  no active
  destination address http https://tools.cisco.com/its/service/oddce/services/DDCEService
  destination address email callhome@cisco.com
  destination transport-method http
  subscribe-to-alert-group diagnostic
  subscribe-to-alert-group environment

subscribe-to-alert-group inventory periodic monthly
  subscribe-to-alert-group configuration periodic monthly
  subscribe-to-alert-group telemetry periodic daily
crashinfo save disable
Cryptochecksum:2e24c7d90262481b8fd7780418f9bfb6
: end

4.2.4 Remote Access IPSEC VPN on Cisco ASA

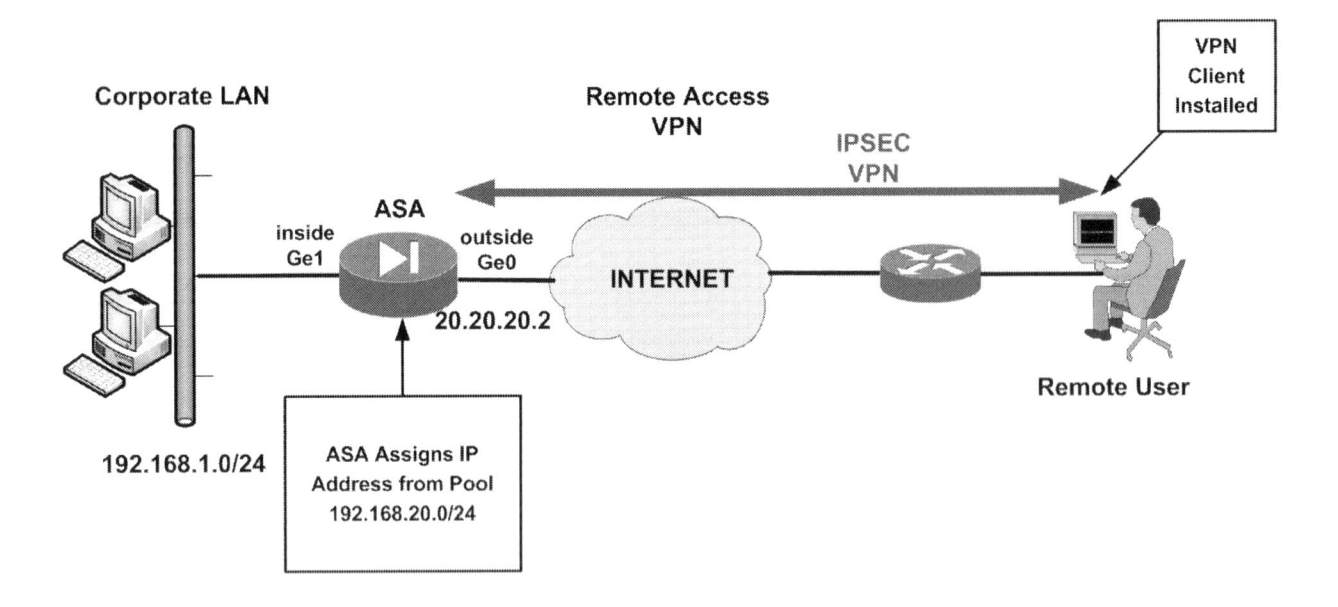

ASA

```
ASA Version 8.4(2)
!
hostname ASA1
enable password 8Ry2YjIyt7RRXU24 encrypted
passwd 2KFQnbNIdI.2KYOU encrypted
names
!
interface GigabitEthernet0
 nameif outside
 security-level 0
 ip address 20.20.20.2 255.255.255.0
!
interface GigabitEthernet1
 nameif inside
 security-level 100
 ip address 192.168.1.1 255.255.255.0
!
ftp mode passive
object network internal_lan
 subnet 192.168.1.0 255.255.255.0
object network obj-local
 subnet 192.168.1.0 255.255.255.0
object network obj-vpnpool
 subnet 192.168.20.0 255.255.255.0
```

```
access-list outside_in extended permit icmp any any echo-reply
access-list outside_in extended deny ip any any log
access-list splittunnel standard permit 192.168.1.0 255.255.255.0
pager lines 24
mtu outside 1500
mtu inside 1500
ip local pool vpnpool 192.168.20.1-192.168.20.254
icmp unreachable rate-limit 1 burst-size 1
no asdm history enable
arp timeout 14400
nat (inside,outside) source static obj-local obj-local destination static obj-vpnpool obj-vpnpool
!
object network internal_lan
 nat (inside,outside) dynamic interface
access-group outside_in in interface outside
route outside 0.0.0.0 0.0.0.0 20.20.20.1 1
timeout xlate 3:00:00
timeout conn 1:00:00 half-closed 0:10:00 udp 0:02:00 icmp 0:00:02
timeout sunrpc 0:10:00 h323 0:05:00 h225 1:00:00 mgcp 0:05:00 mgcp-pat 0:05:00
timeout sip 0:30:00 sip_media 0:02:00 sip-invite 0:03:00 sip-disconnect 0:02:00
timeout sip-provisional-media 0:02:00 uauth 0:05:00 absolute
timeout tcp-proxy-reassembly 0:01:00
timeout floating-conn 0:00:00
dynamic-access-policy-record DfltAccessPolicy
user-identity default-domain LOCAL
no snmp-server location
no snmp-server contact
snmp-server enable traps snmp authentication linkup linkdown coldstart warmstart
crypto ipsec ikev1 transform-set TRSET esp-3des esp-md5-hmac
crypto dynamic-map outside_dyn_map 10 set ikev1 transform-set TRSET
crypto map IPSEC 10 ipsec-isakmp dynamic outside_dyn_map
crypto map IPSEC interface outside
crypto isakmp identity address
crypto ikev1 enable outside
crypto ikev1 policy 10
 authentication pre-share
 encryption 3des
 hash sha
 group 2
 lifetime 86400
telnet timeout 5
ssh timeout 5
console timeout 0
threat-detection basic-threat
threat-detection statistics access-list
no threat-detection statistics tcp-intercept
```

```
group-policy vpn-clients-policy internal
group-policy vpn-clients-policy attributes
 vpn-idle-timeout 30
 dns-server value 192.168.1.5
 split-tunnel-policy tunnelspecified
 split-tunnel-network-list value splittunnel
username vpnuser password test123
tunnel-group remotevpn type remote-access
tunnel-group remotevpn general-attributes
 address-pool vpnpool
 default-group-policy vpn-clients-policy
tunnel-group remotevpn ipsec-attributes
 ikev1 pre-shared-key cisco123
!
!
prompt hostname context
no call-home reporting anonymous
call-home
 profile CiscoTAC-1
  no active
  destination address http https://tools.cisco.com/its/service/oddce/services/DDCEService
  destination address email callhome@cisco.com
  destination transport-method http
  subscribe-to-alert-group diagnostic
  subscribe-to-alert-group environment
  subscribe-to-alert-group inventory periodic monthly
  subscribe-to-alert-group configuration periodic monthly
  subscribe-to-alert-group telemetry periodic daily
crashinfo save disable
Cryptochecksum:9f55a108e0a2df928442f38aac91ec8d
: end
```

4.2.5 Hub-and-Spoke and Remote Access VPN on same device

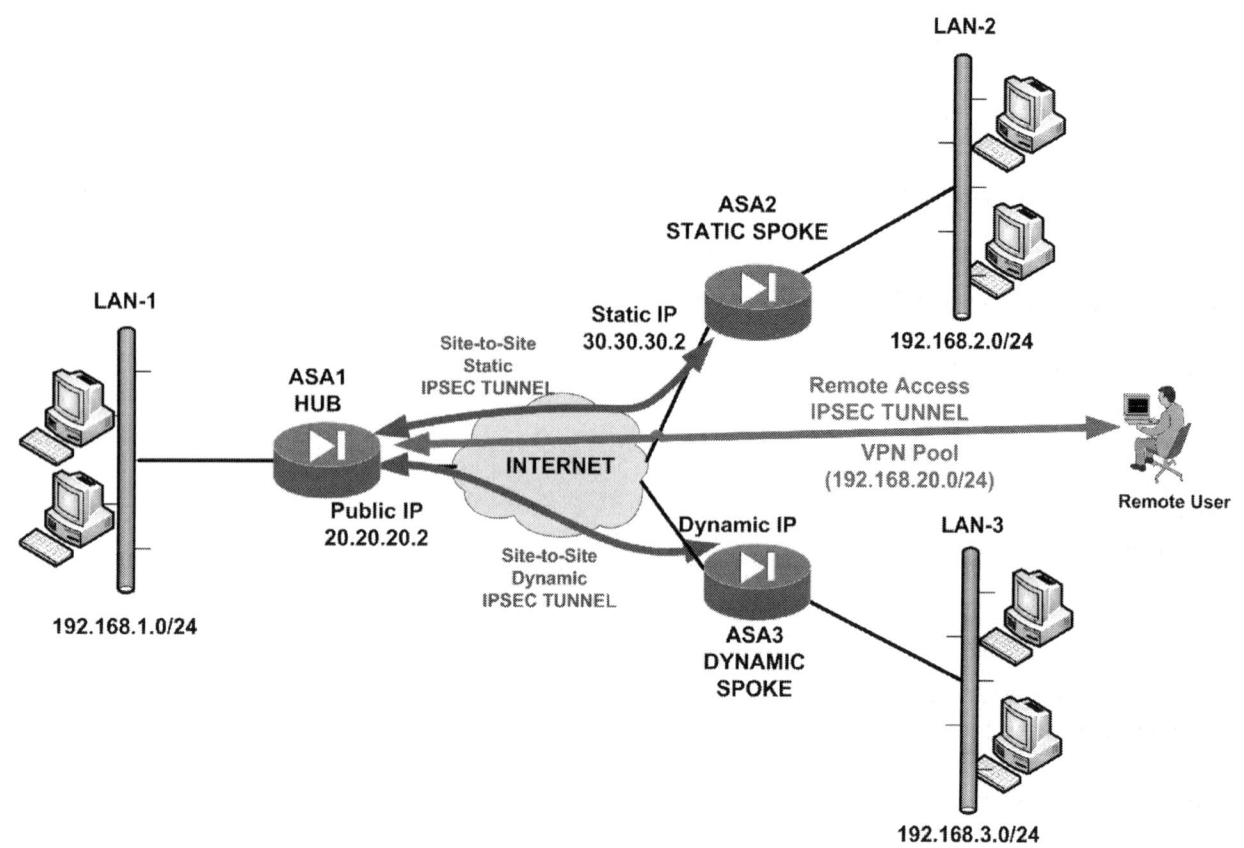

ASA1 (HUB)

ASA Version 8.4(2)
!
hostname ASA1
enable password 8Ry2YjIyt7RRXU24 encrypted
passwd 2KFQnbNIdI.2KYOU encrypted
names
!
interface GigabitEthernet0
 nameif outside
 security-level 0
 ip address 20.20.20.2 255.255.255.0
!
interface GigabitEthernet1
 nameif inside
 security-level 100
 ip address 192.168.1.1 255.255.255.0
!
ftp mode passive

231

```
object network internal_lan
 subnet 192.168.1.0 255.255.255.0
object network obj-local
 subnet 192.168.1.0 255.255.255.0
object network obj-vpnpool
 subnet 192.168.20.0 255.255.255.0
object network obj-remote1
 subnet 192.168.2.0 255.255.255.0
object network obj-remote2
 subnet 192.168.3.0 255.255.255.0
access-list outside_in extended permit icmp any any echo-reply
access-list outside_in extended deny ip any any log
access-list splittunnel standard permit 192.168.1.0 255.255.255.0
access-list VPN-ACL1 extended permit ip 192.168.1.0 255.255.255.0 192.168.2.0 255.255.255.0
access-list VPN-ACL2 extended permit ip 192.168.1.0 255.255.255.0 192.168.3.0 255.255.255.0
pager lines 24
mtu outside 1500
mtu inside 1500
ip local pool vpnpool 192.168.20.1-192.168.20.254
icmp unreachable rate-limit 1 burst-size 1
no asdm history enable
arp timeout 14400
nat (inside,outside) 1 source static obj-local obj-local destination static obj-vpnpool obj-vpnpool no-proxy-arp route-lookup
nat (inside,outside) 2 source static obj-local obj-local destination static obj-remote1 obj-remote1
nat (inside,outside) 3 source static obj-local obj-local destination static obj-remote2 obj-remote2
!
object network internal_lan
 nat (inside,outside) dynamic interface
access-group outside_in in interface outside
route outside 0.0.0.0 0.0.0.0 20.20.20.1 1
timeout xlate 3:00:00
timeout conn 1:00:00 half-closed 0:10:00 udp 0:02:00 icmp 0:00:02
timeout sunrpc 0:10:00 h323 0:05:00 h225 1:00:00 mgcp 0:05:00 mgcp-pat 0:05:00
timeout sip 0:30:00 sip_media 0:02:00 sip-invite 0:03:00 sip-disconnect 0:02:00
timeout sip-provisional-media 0:02:00 uauth 0:05:00 absolute
timeout tcp-proxy-reassembly 0:01:00
timeout floating-conn 0:00:00
dynamic-access-policy-record DfltAccessPolicy
user-identity default-domain LOCAL
no snmp-server location
no snmp-server contact
snmp-server enable traps snmp authentication linkup linkdown coldstart warmstart
crypto ipsec ikev1 transform-set TRSET esp-3des esp-md5-hmac
crypto dynamic-map DYNMAP 5 match address VPN-ACL2
crypto dynamic-map DYNMAP 5 set ikev1 transform-set TRSET
```

```
crypto dynamic-map DYNMAP 10 set ikev1 transform-set TRSET
crypto map IPSEC 5 match address VPN-ACL1
crypto map IPSEC 5 set peer 30.30.30.2
crypto map IPSEC 5 set ikev1 transform-set TRSET
crypto map IPSEC 10 ipsec-isakmp dynamic DYNMAP
crypto map IPSEC interface outside
crypto isakmp identity address
crypto ikev1 enable outside
crypto ikev1 policy 10
 authentication pre-share
 encryption 3des
 hash sha
 group 2
 lifetime 86400
telnet timeout 5
ssh timeout 5
console timeout 0
threat-detection basic-threat
threat-detection statistics access-list
no threat-detection statistics tcp-intercept
group-policy vpn-clients-policy internal
group-policy vpn-clients-policy attributes
 vpn-idle-timeout 30
 split-tunnel-policy tunnelspecified
 split-tunnel-network-list value splittunnel
username vpnuser password test123
tunnel-group remotevpn type remote-access
tunnel-group remotevpn general-attributes
 address-pool vpnpool
 default-group-policy vpn-clients-policy
tunnel-group remotevpn ipsec-attributes
 ikev1 pre-shared-key cisco123
tunnel-group DefaultL2LGroup ipsec-attributes
 ikev1 pre-shared-key secretkey2
tunnel-group 30.30.30.2 type ipsec-l2l
tunnel-group 30.30.30.2 ipsec-attributes
 ikev1 pre-shared-key secretkey1
!
prompt hostname context
no call-home reporting anonymous
call-home
 profile CiscoTAC-1
  no active
  destination address http https://tools.cisco.com/its/service/oddce/services/DDCEService
  destination address email callhome@cisco.com
  destination transport-method http
```

subscribe-to-alert-group diagnostic
  subscribe-to-alert-group environment
  subscribe-to-alert-group inventory periodic monthly
  subscribe-to-alert-group configuration periodic monthly
  subscribe-to-alert-group telemetry periodic daily
crashinfo save disable

ASA2 (Static Spoke)

ASA Version 8.4(2)
!
hostname ASA2
enable password 8Ry2Yjlyt7RRXU24 encrypted
passwd 2KFQnbNIdI.2KYOU encrypted
names
!
interface GigabitEthernet0
 nameif outside
 security-level 0
 ip address 30.30.30.2 255.255.255.0
!
interface GigabitEthernet1
 nameif inside
 security-level 100
 ip address 192.168.2.1 255.255.255.0
!
ftp mode passive
object network obj-local
 subnet 192.168.2.0 255.255.255.0
object network obj-remote
 subnet 192.168.1.0 255.255.255.0
object network internal-lan
 subnet 192.168.2.0 255.255.255.0
access-list outside_in extended permit icmp any any echo-reply
access-list outside_in extended deny ip any any log
access-list VPN-ACL extended permit ip 192.168.2.0 255.255.255.0 192.168.1.0 255.255.255.0
pager lines 24
mtu outside 1500
mtu inside 1500
icmp unreachable rate-limit 1 burst-size 1
no asdm history enable
arp timeout 14400
nat (inside,outside) source static obj-local obj-local destination static obj-remote obj-remote
!

```
object network internal-lan
 nat (inside,outside) dynamic interface
access-group outside_in in interface outside
route outside 0.0.0.0 0.0.0.0 30.30.30.1 1
timeout xlate 3:00:00
timeout conn 1:00:00 half-closed 0:10:00 udp 0:02:00 icmp 0:00:02
timeout sunrpc 0:10:00 h323 0:05:00 h225 1:00:00 mgcp 0:05:00 mgcp-pat 0:05:00
timeout sip 0:30:00 sip_media 0:02:00 sip-invite 0:03:00 sip-disconnect 0:02:00
timeout sip-provisional-media 0:02:00 uauth 0:05:00 absolute
timeout tcp-proxy-reassembly 0:01:00
timeout floating-conn 0:00:00
dynamic-access-policy-record DfltAccessPolicy
user-identity default-domain LOCAL
no snmp-server location
no snmp-server contact
snmp-server enable traps snmp authentication linkup linkdown coldstart warmstart
crypto ipsec ikev1 transform-set TRSET esp-3des esp-md5-hmac
crypto map VPNMAP 5 match address VPN-ACL
crypto map VPNMAP 5 set peer 20.20.20.2
crypto map VPNMAP 5 set ikev1 transform-set TRSET
crypto map VPNMAP interface outside
crypto isakmp identity address
crypto ikev1 enable outside
crypto ikev1 policy 10
 authentication pre-share
 encryption 3des
 hash sha
 group 2
 lifetime 86400
telnet timeout 5
ssh timeout 5
console timeout 0
threat-detection basic-threat
threat-detection statistics access-list
no threat-detection statistics tcp-intercept
tunnel-group 20.20.20.2 type ipsec-l2l
tunnel-group 20.20.20.2 ipsec-attributes
 ikev1 pre-shared-key secretkey1
!
!
prompt hostname context
no call-home reporting anonymous
call-home
 profile CiscoTAC-1
  no active
  destination address http https://tools.cisco.com/its/service/oddce/services/DDCEService
```

destination address email callhome@cisco.com
  destination transport-method http
  subscribe-to-alert-group diagnostic
  subscribe-to-alert-group environment
  subscribe-to-alert-group inventory periodic monthly
  subscribe-to-alert-group configuration periodic monthly
  subscribe-to-alert-group telemetry periodic daily
crashinfo save disable
Cryptochecksum:2e24c7d90262481b8fd7780418f9bfb6
: end

ASA3 (Dynamic Spoke)

ASA Version 8.4(2)
!
hostname ASA3
enable password 8Ry2YjIyt7RRXU24 encrypted
passwd 2KFQnbNIdI.2KYOU encrypted
names
!
interface GigabitEthernet0
 nameif outside
 security-level 0
 ip address dhcp setroute
!
interface GigabitEthernet1
 nameif inside
 security-level 100
 ip address 192.168.3.1 255.255.255.0
!
ftp mode passive
object network obj-local
 subnet 192.168.3.0 255.255.255.0
object network obj-remote
 subnet 192.168.1.0 255.255.255.0
object network internal-lan
 subnet 192.168.3.0 255.255.255.0
access-list outside_in extended permit icmp any any echo-reply
access-list outside_in extended deny ip any any log
access-list VPN-ACL extended permit ip 192.168.3.0 255.255.255.0 192.168.1.0 255.255.255.0
pager lines 24
mtu outside 1500
mtu inside 1500
icmp unreachable rate-limit 1 burst-size 1

no asdm history enable
arp timeout 14400
nat (inside,outside) source static obj-local obj-local destination static obj-remote obj-remote
!
object network internal-lan
 nat (inside,outside) dynamic interface
access-group outside_in in interface outside
timeout xlate 3:00:00
timeout conn 1:00:00 half-closed 0:10:00 udp 0:02:00 icmp 0:00:02
timeout sunrpc 0:10:00 h323 0:05:00 h225 1:00:00 mgcp 0:05:00 mgcp-pat 0:05:00
timeout sip 0:30:00 sip_media 0:02:00 sip-invite 0:03:00 sip-disconnect 0:02:00
timeout sip-provisional-media 0:02:00 uauth 0:05:00 absolute
timeout tcp-proxy-reassembly 0:01:00
timeout floating-conn 0:00:00
dynamic-access-policy-record DfltAccessPolicy
user-identity default-domain LOCAL
no snmp-server location
no snmp-server contact
snmp-server enable traps snmp authentication linkup linkdown coldstart warmstart
crypto ipsec ikev1 transform-set TRSET esp-3des esp-md5-hmac
crypto map VPNMAP 5 match address VPN-ACL
crypto map VPNMAP 5 set peer 20.20.20.2
crypto map VPNMAP 5 set ikev1 transform-set TRSET
crypto map VPNMAP interface outside
crypto isakmp identity address
crypto ikev1 enable outside
crypto ikev1 policy 10
 authentication pre-share
 encryption 3des
 hash sha
 group 2
 lifetime 86400
telnet timeout 5
ssh timeout 5
console timeout 0
threat-detection basic-threat
threat-detection statistics access-list
no threat-detection statistics tcp-intercept
tunnel-group 20.20.20.2 type ipsec-l2l
tunnel-group 20.20.20.2 ipsec-attributes
 ikev1 pre-shared-key secretkey2
!
!
prompt hostname context
no call-home reporting anonymous
call-home

profile CiscoTAC-1
  no active
  destination address http https://tools.cisco.com/its/service/oddce/services/DDCEService
  destination address email callhome@cisco.com
  destination transport-method http
  subscribe-to-alert-group diagnostic
  subscribe-to-alert-group environment
  subscribe-to-alert-group inventory periodic monthly
  subscribe-to-alert-group configuration periodic monthly
  subscribe-to-alert-group telemetry periodic daily
crashinfo save disable
Cryptochecksum:2e24c7d90262481b8fd7780418f9bfb6
: end

4.2.6 Site-to-Site IPSEC VPN with failover using backup ISP

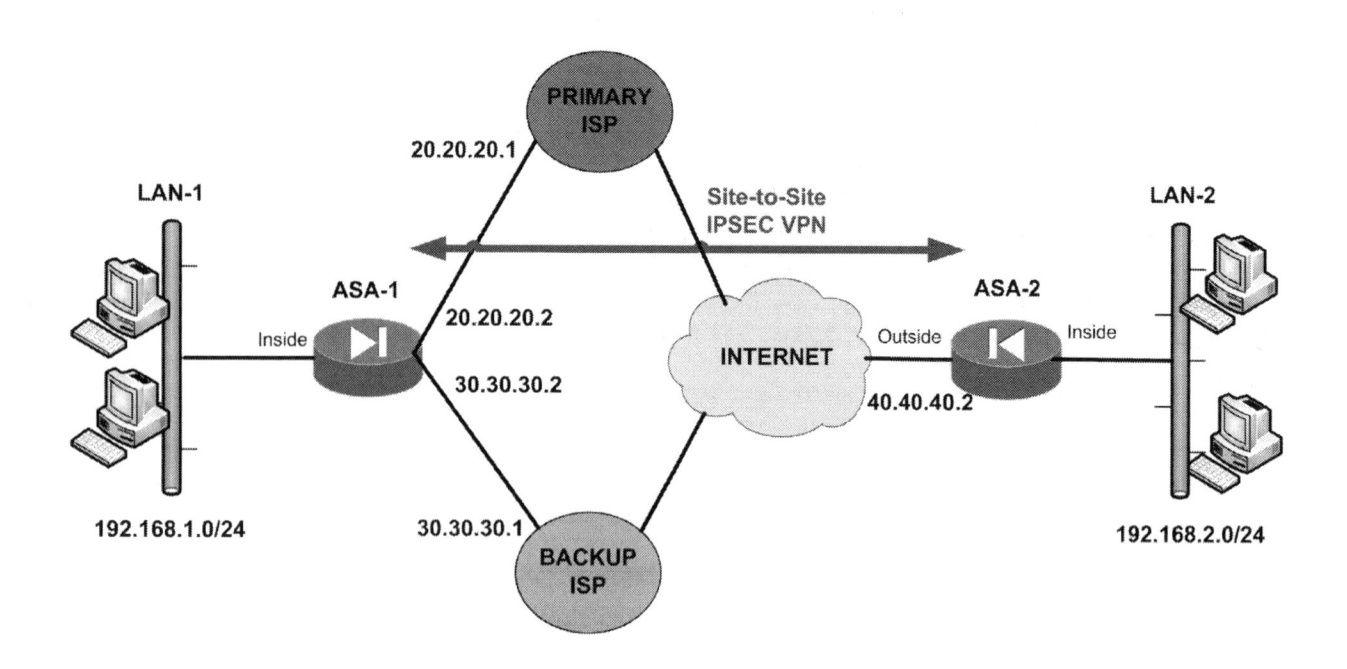

ASA-1

```
ASA Version 8.4(2)
!
hostname ASA1
enable password 8Ry2Yjlyt7RRXU24 encrypted
passwd 2KFQnbNIdI.2KYOU encrypted
names
!
interface GigabitEthernet0
 nameif primary-isp
 security-level 0
 ip address 20.20.20.2 255.255.255.0
!
interface GigabitEthernet1
 nameif backup-isp
 security-level 0
 ip address 30.30.30.2 255.255.255.0
!
interface GigabitEthernet2
 nameif inside
 security-level 100
 ip address 192.168.1.1 255.255.255.0
!
```

ftp mode passive

object network obj-local
 subnet 192.168.1.0 255.255.255.0
object network obj-remote
 subnet 192.168.2.0 255.255.255.0
object network internal-lan
 subnet 192.168.1.0 255.255.255.0
access-list outside_in extended permit icmp any any echo-reply
access-list outside_in extended deny ip any any log
access-list VPN-ACL extended permit ip 192.168.1.0 255.255.255.0 192.168.2.0 255.255.255.0
pager lines 24
mtu primary-isp 1500
mtu backup-isp 1500
mtu inside 1500
icmp unreachable rate-limit 1 burst-size 1
no asdm history enable
arp timeout 14400
nat (inside,primary-isp) source static obj-local obj-local destination static obj-remote obj-remote
nat (inside,backup-isp) source static obj-local obj-local destination static obj-remote obj-remote
access-group outside_in in interface primary-isp
access-group outside_in in interface backup-isp
route primary-isp 0.0.0.0 0.0.0.0 20.20.20.1 1 track 10
route backup-isp 0.0.0.0 0.0.0.0 30.30.30.1 254
timeout xlate 3:00:00
timeout conn 1:00:00 half-closed 0:10:00 udp 0:02:00 icmp 0:00:02
timeout sunrpc 0:10:00 h323 0:05:00 h225 1:00:00 mgcp 0:05:00 mgcp-pat 0:05:00
timeout sip 0:30:00 sip_media 0:02:00 sip-invite 0:03:00 sip-disconnect 0:02:00
timeout sip-provisional-media 0:02:00 uauth 0:05:00 absolute
timeout tcp-proxy-reassembly 0:01:00
timeout floating-conn 0:00:00
dynamic-access-policy-record DfltAccessPolicy
user-identity default-domain LOCAL
no snmp-server location
no snmp-server contact
snmp-server enable traps snmp authentication linkup linkdown coldstart warmstart
sla monitor 100
 type echo protocol ipIcmpEcho 20.20.20.1 interface primary-isp
 timeout 3000
 frequency 5
sla monitor schedule 100 life forever start-time now
crypto ipsec ikev1 transform-set TRSET esp-3des esp-md5-hmac
crypto map VPNMAP 10 match address VPN-ACL
crypto map VPNMAP 10 set peer 40.40.40.2
crypto map VPNMAP 10 set ikev1 transform-set TRSET
crypto map VPNMAP interface primary-isp

```
 crypto map VPNMAP interface backup-isp
 crypto isakmp identity address
 crypto ikev1 enable primary-isp
 crypto ikev1 enable backup-isp
 crypto ikev1 policy 10
  authentication pre-share
  encryption 3des
  hash sha
  group 2
  lifetime 86400
 !
 track 10 rtr 100 reachability
 telnet timeout 5
 ssh timeout 5
 console timeout 0
 threat-detection basic-threat
 threat-detection statistics access-list
 no threat-detection statistics tcp-intercept
 tunnel-group 40.40.40.2 type ipsec-l2l
 tunnel-group 40.40.40.2 ipsec-attributes
  ikev1 pre-shared-key secretkey1
 !
 !
 prompt hostname context
 no call-home reporting anonymous
 call-home
  profile CiscoTAC-1
   no active
   destination address http https://tools.cisco.com/its/service/oddce/services/DDCEService
   destination address email callhome@cisco.com
   destination transport-method http
   subscribe-to-alert-group diagnostic
   subscribe-to-alert-group environment
   subscribe-to-alert-group inventory periodic monthly
   subscribe-to-alert-group configuration periodic monthly
   subscribe-to-alert-group telemetry periodic daily
 crashinfo save disable
 Cryptochecksum:d4e5496553edfa8a692219b5d5ef245c
 : end
```

ASA-2

```
ASA Version 8.4(2)
!
hostname ASA2
enable password 8Ry2YjIyt7RRXU24 encrypted
passwd 2KFQnbNIdI.2KYOU encrypted
names
!
interface GigabitEthernet0
 nameif outside
 security-level 0
 ip address 40.40.40.2 255.255.255.0
!
interface GigabitEthernet1
 nameif inside
 security-level 100
 ip address 192.168.2.1 255.255.255.0
!
ftp mode passive
object network obj-local
 subnet 192.168.2.0 255.255.255.0
object network obj-remote
 subnet 192.168.1.0 255.255.255.0
object network internal-lan
 subnet 192.168.2.0 255.255.255.0
access-list outside_in extended permit icmp any any echo-reply
access-list outside_in extended deny ip any any log
access-list VPN-ACL extended permit ip 192.168.2.0 255.255.255.0 192.168.1.0 255.255.255.0
pager lines 24
mtu outside 1500
mtu inside 1500
icmp unreachable rate-limit 1 burst-size 1
no asdm history enable
arp timeout 14400
nat (inside,outside) source static obj-local obj-local destination static obj-remote obj-remote
!
object network internal-lan
 nat (inside,outside) dynamic interface
access-group outside_in in interface outside
route outside 0.0.0.0 0.0.0.0 40.40.40.1 1
timeout xlate 3:00:00
timeout conn 1:00:00 half-closed 0:10:00 udp 0:02:00 icmp 0:00:02
timeout sunrpc 0:10:00 h323 0:05:00 h225 1:00:00 mgcp 0:05:00 mgcp-pat 0:05:00
timeout sip 0:30:00 sip_media 0:02:00 sip-invite 0:03:00 sip-disconnect 0:02:00
```

timeout sip-provisional-media 0:02:00 uauth 0:05:00 absolute
 timeout tcp-proxy-reassembly 0:01:00
 timeout floating-conn 0:00:00
 dynamic-access-policy-record DfltAccessPolicy
 user-identity default-domain LOCAL
 no snmp-server location
 no snmp-server contact
 snmp-server enable traps snmp authentication linkup linkdown coldstart warmstart
 crypto ipsec ikev1 transform-set TRSET esp-3des esp-md5-hmac
 crypto map VPNMAP 10 match address VPN-ACL
 crypto map VPNMAP 10 set peer 20.20.20.2 30.30.30.2
 crypto map VPNMAP 10 set ikev1 transform-set TRSET
 crypto map VPNMAP interface outside
 crypto isakmp identity address
 crypto ikev1 enable outside
 crypto ikev1 policy 10
  authentication pre-share
  encryption 3des
  hash sha
  group 2
  lifetime 86400
 telnet timeout 5
 ssh timeout 5
 console timeout 0
 threat-detection basic-threat
 threat-detection statistics access-list
 no threat-detection statistics tcp-intercept
 tunnel-group 20.20.20.2 type ipsec-l2l
 tunnel-group 20.20.20.2 ipsec-attributes
  ikev1 pre-shared-key secretkey1
 tunnel-group 30.30.30.2 type ipsec-l2l
 tunnel-group 30.30.30.2 ipsec-attributes
  ikev1 pre-shared-key secretkey1
 !
 !
 prompt hostname context
 no call-home reporting anonymous
 call-home
  profile CiscoTAC-1
   no active
   destination address http https://tools.cisco.com/its/service/oddce/services/DDCEService
   destination address email callhome@cisco.com
   destination transport-method http
   subscribe-to-alert-group diagnostic
   subscribe-to-alert-group environment
   subscribe-to-alert-group inventory periodic monthly

subscribe-to-alert-group configuration periodic monthly
  subscribe-to-alert-group telemetry periodic daily
crashinfo save disable
Cryptochecksum:57e963c9585371ad20093abffad1eb5c
: end

4.2.7 Site-to-Site IPSEC VPN with Duplicate Subnets-Example1

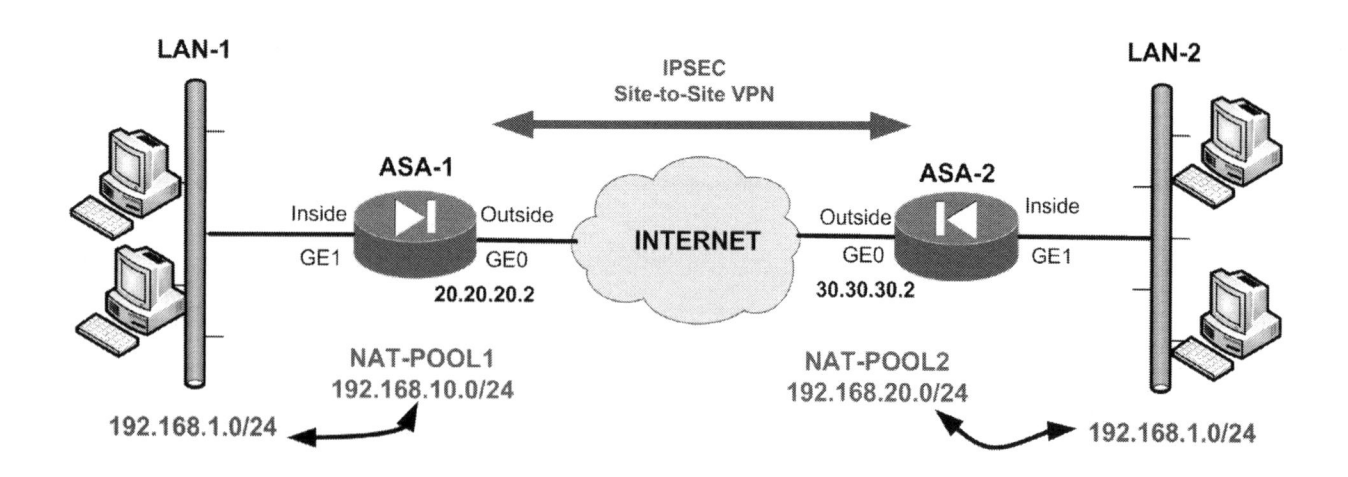

ASA-1

```
ASA Version 8.4(2)
!
hostname ASA1
enable password 8Ry2YjIyt7RRXU24 encrypted
passwd 2KFQnbNIdI.2KYOU encrypted
names
!
interface GigabitEthernet0
 nameif outside
 security-level 0
 ip address 20.20.20.2 255.255.255.0
!
interface GigabitEthernet1
 nameif inside
 security-level 100
 ip address 192.168.1.1 255.255.255.0
!
ftp mode passive
object network obj-local
 subnet 192.168.1.0 255.255.255.0
object network NAT-POOL1
 subnet 192.168.10.0 255.255.255.0
object network NAT-POOL2
 subnet 192.168.20.0 255.255.255.0

access-list outside_in extended permit icmp any any echo-reply
access-list outside_in extended deny ip any any log
```

access-list VPN-ACL extended permit ip 192.168.10.0 255.255.255.0 192.168.20.0 255.255.255.0

pager lines 24
mtu outside 1500
mtu inside 1500
icmp unreachable rate-limit 1 burst-size 1
no asdm history enable
arp timeout 14400
nat (inside,outside) source static obj-local NAT-POOL1 destination static NAT-POOL2 NAT-POOL2
!
object network obj-local
 nat (inside,outside) dynamic interface

access-group outside_in in interface outside

route outside 0.0.0.0 0.0.0.0 20.20.20.1 1
timeout xlate 3:00:00
timeout conn 1:00:00 half-closed 0:10:00 udp 0:02:00 icmp 0:00:02
timeout sunrpc 0:10:00 h323 0:05:00 h225 1:00:00 mgcp 0:05:00 mgcp-pat 0:05:00
timeout sip 0:30:00 sip_media 0:02:00 sip-invite 0:03:00 sip-disconnect 0:02:00
timeout sip-provisional-media 0:02:00 uauth 0:05:00 absolute
timeout tcp-proxy-reassembly 0:01:00
timeout floating-conn 0:00:00
dynamic-access-policy-record DfltAccessPolicy
user-identity default-domain LOCAL
no snmp-server location
no snmp-server contact
snmp-server enable traps snmp authentication linkup linkdown coldstart warmstart
crypto ipsec ikev1 transform-set TRSET esp-3des esp-md5-hmac
crypto map VPNMAP 10 match address VPN-ACL
crypto map VPNMAP 10 set peer 30.30.30.2
crypto map VPNMAP 10 set ikev1 transform-set TRSET
crypto map VPNMAP interface outside
crypto isakmp identity address
crypto ikev1 enable outside
crypto ikev1 policy 10
 authentication pre-share
 encryption 3des
 hash sha
 group 2
 lifetime 86400
telnet timeout 5
ssh timeout 5
console timeout 0
threat-detection basic-threat
threat-detection statistics access-list

```
 no threat-detection statistics tcp-intercept
tunnel-group 30.30.30.2 type ipsec-l2l
tunnel-group 30.30.30.2 ipsec-attributes
 ikev1 pre-shared-key testkey123
!
!
prompt hostname context
no call-home reporting anonymous
call-home
 profile CiscoTAC-1
  no active
  destination address http https://tools.cisco.com/its/service/oddce/services/DDCEService
  destination address email callhome@cisco.com
  destination transport-method http
  subscribe-to-alert-group diagnostic
  subscribe-to-alert-group environment
  subscribe-to-alert-group inventory periodic monthly
  subscribe-to-alert-group configuration periodic monthly
  subscribe-to-alert-group telemetry periodic daily
crashinfo save disable
Cryptochecksum:2e24c7d90262481b8fd7780418f9bfb6
: end
```

ASA-2

```
ASA Version 8.4(2)
!
hostname ASA2
enable password 8Ry2YjIyt7RRXU24 encrypted
passwd 2KFQnbNIdI.2KYOU encrypted
names
!
interface GigabitEthernet0
 nameif outside
 security-level 0
 ip address 30.30.30.2 255.255.255.0
!
interface GigabitEthernet1
 nameif inside
 security-level 100
 ip address 192.168.1.1 255.255.255.0
!
ftp mode passive
object network obj-local
 subnet 192.168.1.0 255.255.255.0
```

```
object network NAT-POOL1
 subnet 192.168.10.0 255.255.255.0
object network NAT-POOL2
 subnet 192.168.20.0 255.255.255.0

access-list outside_in extended permit icmp any any echo-reply
access-list outside_in extended deny ip any any log

access-list VPN-ACL extended permit ip 192.168.20.0 255.255.255.0 192.168.10.0 255.255.255.0

pager lines 24
mtu outside 1500
mtu inside 1500
icmp unreachable rate-limit 1 burst-size 1
no asdm history enable
arp timeout 14400
nat (inside,outside) source static obj-local NAT-POOL2 destination static NAT-POOL1 NAT-POOL1
!
object network obj-local
 nat (inside,outside) dynamic interface

access-group outside_in in interface outside

route outside 0.0.0.0 0.0.0.0 30.30.30.1 1
timeout xlate 3:00:00
timeout conn 1:00:00 half-closed 0:10:00 udp 0:02:00 icmp 0:00:02
timeout sunrpc 0:10:00 h323 0:05:00 h225 1:00:00 mgcp 0:05:00 mgcp-pat 0:05:00
timeout sip 0:30:00 sip_media 0:02:00 sip-invite 0:03:00 sip-disconnect 0:02:00
timeout sip-provisional-media 0:02:00 uauth 0:05:00 absolute
timeout tcp-proxy-reassembly 0:01:00
timeout floating-conn 0:00:00
dynamic-access-policy-record DfltAccessPolicy
user-identity default-domain LOCAL
no snmp-server location
no snmp-server contact
snmp-server enable traps snmp authentication linkup linkdown coldstart warmstart
crypto ipsec ikev1 transform-set TRSET esp-3des esp-md5-hmac
crypto map VPNMAP 10 match address VPN-ACL
crypto map VPNMAP 10 set peer 20.20.20.2
crypto map VPNMAP 10 set ikev1 transform-set TRSET
crypto map VPNMAP interface outside
crypto isakmp identity address
crypto ikev1 enable outside
crypto ikev1 policy 10
 authentication pre-share
 encryption 3des
```

hash sha
  group 2
  lifetime 86400
 telnet timeout 5
 ssh timeout 5
 console timeout 0
 threat-detection basic-threat
 threat-detection statistics access-list
 no threat-detection statistics tcp-intercept
 tunnel-group 20.20.20.2 type ipsec-l2l
 tunnel-group 20.20.20.2 ipsec-attributes
  ikev1 pre-shared-key testkey123
 !
 !
 prompt hostname context
 no call-home reporting anonymous
 call-home
  profile CiscoTAC-1
   no active
   destination address http https://tools.cisco.com/its/service/oddce/services/DDCEService
   destination address email callhome@cisco.com
   destination transport-method http
   subscribe-to-alert-group diagnostic
   subscribe-to-alert-group environment
   subscribe-to-alert-group inventory periodic monthly
   subscribe-to-alert-group configuration periodic monthly
   subscribe-to-alert-group telemetry periodic daily
 crashinfo save disable
 Cryptochecksum:2e24c7d90262481b8fd7780418f9bfb6
 : end

4.2.8 Site-to-Site IPSEC VPN with Duplicate Subnets-Example2

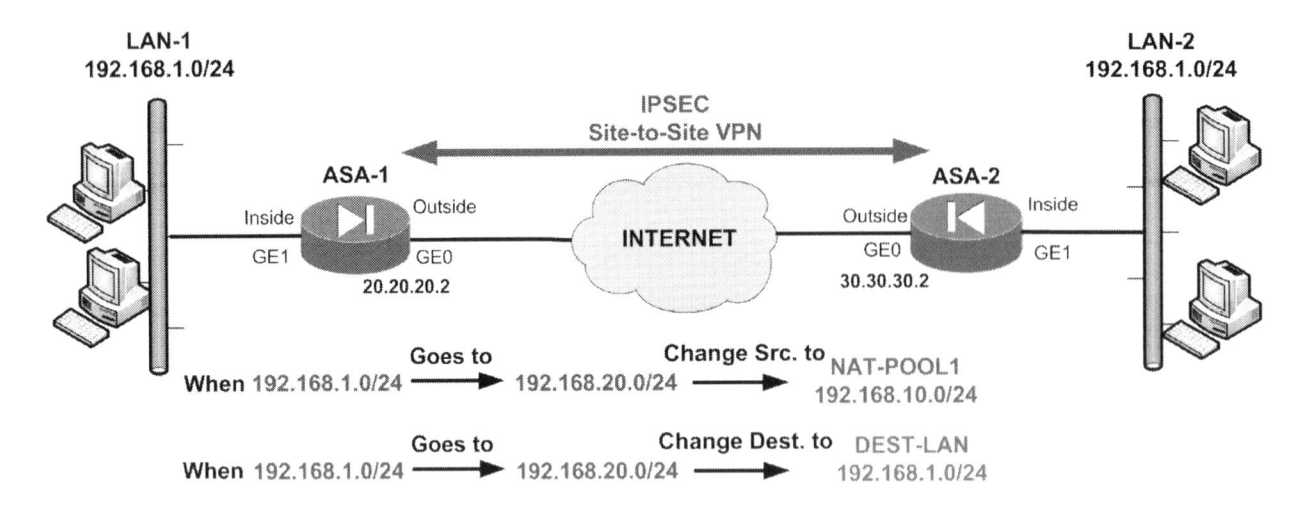

LAN-1 192.168.1.0/24

LAN-2 192.168.1.0/24

IPSEC Site-to-Site VPN

ASA-1 — Inside GE1 / Outside GE0 — 20.20.20.2

INTERNET

ASA-2 — Outside GE0 / Inside GE1 — 30.30.30.2

When 192.168.1.0/24 — Goes to → 192.168.20.0/24 — Change Src. to NAT-POOL1 192.168.10.0/24

When 192.168.1.0/24 — Goes to → 192.168.20.0/24 — Change Dest. to DEST-LAN 192.168.1.0/24

ASA-1

```
ASA Version 8.4(2)
!
hostname ASA1
enable password 8Ry2YjIyt7RRXU24 encrypted
passwd 2KFQnbNIdI.2KYOU encrypted
names
!
interface GigabitEthernet0
 nameif outside
 security-level 0
 ip address 20.20.20.2 255.255.255.0
!
interface GigabitEthernet1
 nameif inside
 security-level 100
 ip address 192.168.1.1 255.255.255.0
!
ftp mode passive
object network LOCAL-LAN
 subnet 192.168.1.0 255.255.255.0
object network NAT-POOL1
 subnet 192.168.10.0 255.255.255.0
object network NAT-POOL2
 subnet 192.168.20.0 255.255.255.0
object network DEST-LAN
 subnet 192.168.1.0 255.255.255.0
```

```
object network inside-lan
 subnet 192.168.1.0 255.255.255.0
access-list outside_in extended permit icmp any any echo-reply
access-list outside_in extended deny ip any any log
access-list VPN-ACL extended permit ip 192.168.10.0 255.255.255.0 192.168.1.0 255.255.255.0
pager lines 24
mtu outside 1500
mtu inside 1500
icmp unreachable rate-limit 1 burst-size 1
no asdm history enable
arp timeout 14400
nat (inside,outside) source static LOCAL-LAN NAT-POOL1 destination static NAT-POOL2 DEST-LAN
!
object network DEST-LAN
 nat (outside,inside) static NAT-POOL2
object network inside-lan
 nat (inside,outside) dynamic interface
access-group outside_in in interface outside
route outside 0.0.0.0 0.0.0.0 20.20.20.1 1
timeout xlate 3:00:00
timeout conn 1:00:00 half-closed 0:10:00 udp 0:02:00 icmp 0:00:02
timeout sunrpc 0:10:00 h323 0:05:00 h225 1:00:00 mgcp 0:05:00 mgcp-pat 0:05:00
timeout sip 0:30:00 sip_media 0:02:00 sip-invite 0:03:00 sip-disconnect 0:02:00
timeout sip-provisional-media 0:02:00 uauth 0:05:00 absolute
timeout tcp-proxy-reassembly 0:01:00
timeout floating-conn 0:00:00
dynamic-access-policy-record DfltAccessPolicy
user-identity default-domain LOCAL
no snmp-server location
no snmp-server contact
snmp-server enable traps snmp authentication linkup linkdown coldstart warmstart
crypto ipsec ikev1 transform-set TRSET esp-3des esp-md5-hmac
crypto map VPNMAP 10 match address VPN-ACL
crypto map VPNMAP 10 set peer 30.30.30.2
crypto map VPNMAP 10 set ikev1 transform-set TRSET
crypto map VPNMAP interface outside
crypto isakmp identity address
crypto ikev1 enable outside
crypto ikev1 policy 10
 authentication pre-share
 encryption 3des
 hash sha
 group 2
 lifetime 86400
telnet timeout 5
ssh timeout 5
```

```
console timeout 0
threat-detection basic-threat
threat-detection statistics access-list
no threat-detection statistics tcp-intercept
tunnel-group 30.30.30.2 type ipsec-l2l
tunnel-group 30.30.30.2 ipsec-attributes
 ikev1 pre-shared-key testkey123
!
!
prompt hostname context
no call-home reporting anonymous
call-home
 profile CiscoTAC-1
  no active
  destination address http https://tools.cisco.com/its/service/oddce/services/DDCEService
  destination address email callhome@cisco.com
  destination transport-method http
  subscribe-to-alert-group diagnostic
  subscribe-to-alert-group environment
  subscribe-to-alert-group inventory periodic monthly
  subscribe-to-alert-group configuration periodic monthly
  subscribe-to-alert-group telemetry periodic daily
crashinfo save disable
Cryptochecksum:d41d8cd98f00b204e9800998ecf8427e
: end
```

ASA-2

```
ASA Version 8.4(2)
!
hostname ASA2
enable password 8Ry2YjIyt7RRXU24 encrypted
passwd 2KFQnbNIdI.2KYOU encrypted
names
!
interface GigabitEthernet0
 nameif outside
 security-level 0
 ip address 30.30.30.2 255.255.255.0
!
interface GigabitEthernet1
 nameif inside
```

```
 security-level 100
 ip address 192.168.1.1 255.255.255.0
 !
ftp mode passive
object network obj-local
 subnet 192.168.1.0 255.255.255.0
object network DEST-LAN
 subnet 192.168.10.0 255.255.255.0
access-list outside_in extended permit icmp any any echo-reply
access-list outside_in extended deny ip any any log
access-list VPN-ACL extended permit ip 192.168.1.0 255.255.255.0 192.168.10.0 255.255.255.0
pager lines 24
mtu outside 1500
mtu inside 1500
icmp unreachable rate-limit 1 burst-size 1
no asdm history enable
arp timeout 14400
nat (inside,outside) source static obj-local obj-local destination static DEST-LAN DEST-LAN
!
object network obj-local
 nat (inside,outside) dynamic interface
access-group outside_in in interface outside
route outside 0.0.0.0 0.0.0.0 30.30.30.1 1
timeout xlate 3:00:00
timeout conn 1:00:00 half-closed 0:10:00 udp 0:02:00 icmp 0:00:02
timeout sunrpc 0:10:00 h323 0:05:00 h225 1:00:00 mgcp 0:05:00 mgcp-pat 0:05:00
timeout sip 0:30:00 sip_media 0:02:00 sip-invite 0:03:00 sip-disconnect 0:02:00
timeout sip-provisional-media 0:02:00 uauth 0:05:00 absolute
timeout tcp-proxy-reassembly 0:01:00
timeout floating-conn 0:00:00
dynamic-access-policy-record DfltAccessPolicy
user-identity default-domain LOCAL
no snmp-server location
no snmp-server contact
snmp-server enable traps snmp authentication linkup linkdown coldstart warmstart
crypto ipsec ikev1 transform-set TRSET esp-3des esp-md5-hmac
crypto map VPNMAP 10 match address VPN-ACL
crypto map VPNMAP 10 set peer 20.20.20.2
crypto map VPNMAP 10 set ikev1 transform-set TRSET
crypto map VPNMAP interface outside
crypto isakmp identity address
crypto ikev1 enable outside
crypto ikev1 policy 10
 authentication pre-share
 encryption 3des
 hash sha
```

group 2
 lifetime 86400
telnet timeout 5
ssh timeout 5
console timeout 0
threat-detection basic-threat
threat-detection statistics access-list
no threat-detection statistics tcp-intercept
tunnel-group 20.20.20.2 type ipsec-l2l
tunnel-group 20.20.20.2 ipsec-attributes
 ikev1 pre-shared-key testkey123
!
!
prompt hostname context
no call-home reporting anonymous
call-home
 profile CiscoTAC-1
  no active
  destination address http https://tools.cisco.com/its/service/oddce/services/DDCEService
  destination address email callhome@cisco.com
  destination transport-method http
  subscribe-to-alert-group diagnostic
  subscribe-to-alert-group environment
  subscribe-to-alert-group inventory periodic monthly
  subscribe-to-alert-group configuration periodic monthly
  subscribe-to-alert-group telemetry periodic daily
crashinfo save disable
Cryptochecksum:c76a3d215333f48da61941fc17bde9e1
: end
ASA2#

4.2.9 Anyconnect SSL Web VPN

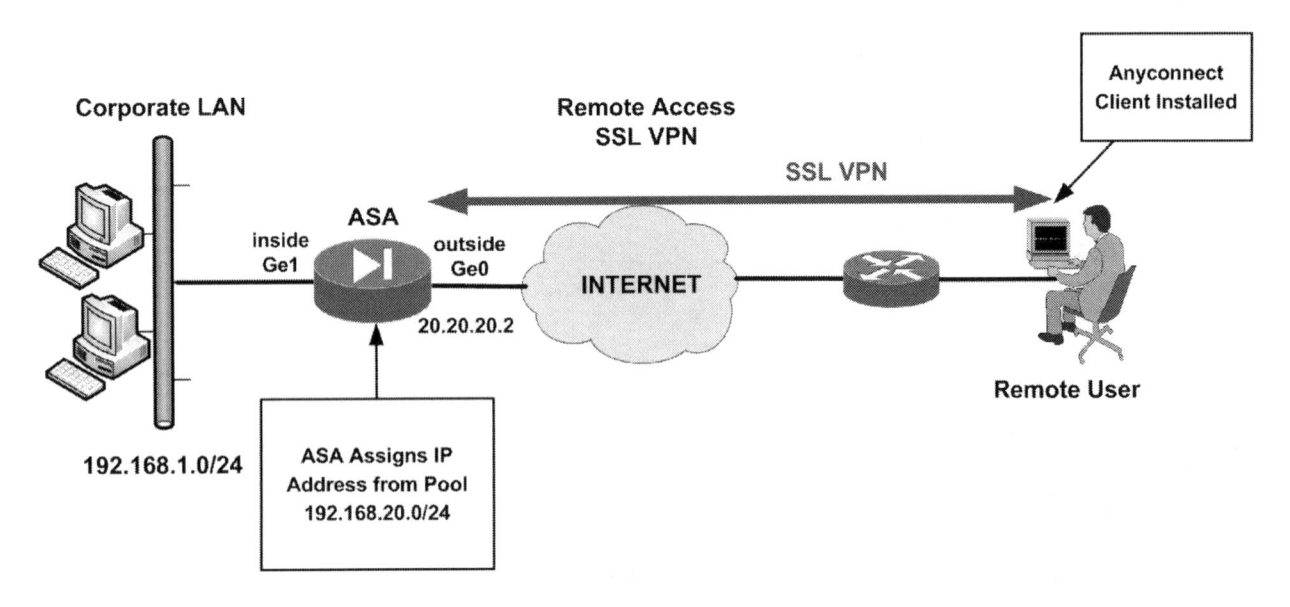

ASA

ASA Version 8.4(2)
!
hostname ASA1
enable password 8Ry2YjIyt7RRXU24 encrypted
passwd 2KFQnbNIdI.2KYOU encrypted
names
!
interface GigabitEthernet0
 nameif outside
 security-level 0
 ip address 20.20.20.2 255.255.255.0
!
interface GigabitEthernet1
 nameif inside
 security-level 100
 ip address 192.168.1.1 255.255.255.0
!
ftp mode passive
object network internal_lan
 subnet 192.168.1.0 255.255.255.0
object network obj-local
 subnet 192.168.1.0 255.255.255.0
object network obj-vpnpool
 subnet 192.168.20.0 255.255.255.0

```
access-list outside_in extended permit icmp any any echo-reply
access-list outside_in extended deny ip any any log
access-list split-tunnel standard permit 192.168.1.0 255.255.255.0
pager lines 24
mtu outside 1500
mtu inside 1500
ip local pool vpnpool 192.168.20.1-192.168.20.254 mask 255.255.255.0
icmp unreachable rate-limit 1 burst-size 1
no asdm history enable
arp timeout 14400
nat (inside,outside) source static obj-local obj-local destination static obj-vpnpool obj-vpnpool no-proxy-arp route-lookup
!
object network internal_lan
 nat (inside,outside) dynamic interface
access-group outside_in in interface outside
route outside 0.0.0.0 0.0.0.0 20.20.20.1 1
timeout xlate 3:00:00
timeout conn 1:00:00 half-closed 0:10:00 udp 0:02:00 icmp 0:00:02
timeout sunrpc 0:10:00 h323 0:05:00 h225 1:00:00 mgcp 0:05:00 mgcp-pat 0:05:00
timeout sip 0:30:00 sip_media 0:02:00 sip-invite 0:03:00 sip-disconnect 0:02:00
timeout sip-provisional-media 0:02:00 uauth 0:05:00 absolute
timeout tcp-proxy-reassembly 0:01:00
timeout floating-conn 0:00:00
dynamic-access-policy-record DfltAccessPolicy
user-identity default-domain LOCAL
no snmp-server location
no snmp-server contact
snmp-server enable traps snmp authentication linkup linkdown coldstart warmstart
telnet timeout 5
ssh timeout 5
console timeout 0
threat-detection basic-threat
threat-detection statistics access-list
no threat-detection statistics tcp-intercept
webvpn
 enable outside
 anyconnect image disk0:/anyconnect-win-2.4.1012-k9.pkg 1
 anyconnect enable
 tunnel-group-list enable
group-policy SSLVPNpolicy internal
group-policy SSLVPNpolicy attributes
 vpn-tunnel-protocol ssl-client
 split-tunnel-policy tunnelspecified
 split-tunnel-network-list value split-tunnel
 webvpn
```

```
  anyconnect keep-installer installed
  anyconnect ask enable default anyconnect timeout 10
username sslvpnuser password test123
tunnel-group SSLVPNprofile type remote-access
tunnel-group SSLVPNprofile general-attributes
 address-pool vpnpool
 default-group-policy SSLVPNpolicy
tunnel-group SSLVPNprofile webvpn-attributes
 group-alias SSL_USERS enable
!
!
prompt hostname context
no call-home reporting anonymous
call-home
 profile CiscoTAC-1
  no active
  destination address http https://tools.cisco.com/its/service/oddce/services/DDCEService
  destination address email callhome@cisco.com
  destination transport-method http
  subscribe-to-alert-group diagnostic
  subscribe-to-alert-group environment
  subscribe-to-alert-group inventory periodic monthly
  subscribe-to-alert-group configuration periodic monthly
  subscribe-to-alert-group telemetry periodic daily
crashinfo save disable
Cryptochecksum:afb950456271f03ac77b5e3387988bf6
: end
```

Conclusion:

If you have studied carefully the information presented in this book, I'm confident that you will be able to tackle the most frequent VPN configuration scenarios that you will encounter in your professional career. I have included all of my knowledge about Cisco VPNs and I hope that I have covered both well known and hard-to-find scenarios.

I know that it's not possible to satisfy all readers of this book with the content I have included. Nevertheless, I believe that this book will be a great reference for every networking professional because a network engineer will certainly encounter a VPN configuration task sometime in his/her career.

Again, thank you for purchasing and reading this book. It has been a pleasure writing this guide, and I really hope that you have enjoyed it as well.

You can check out my Networking related Blog http://www.networkstraining.com for more technical tips and tutorials about Cisco products and solutions. You can also subscribe with your email address at my Blog above in order to receive news and updates about my books and other Cisco technical tips and tutorials.

If you are interested in my other book "**Cisco ASA Firewall Fundamentals-3rd Edition**", you can check it out here: http://www.networkstraining.com/ciscoasaebook.php

I will be glad to answer any questions you may have at admin@networkstraining.com

GOOD LUCK TO YOUR PROFESSIONAL CAREER

Index:

AAA7, 53, 54, 161, 162, 163, 164, 165, 166, 167
ACL 95
AnyConnect3, 30, 31, 32, 151, 152, 154, 155, 157, 160
asymmetrical authentication 149
Authentication Header 13
backup ISP 4, 6, 128, 129, 131, 251
Checkpoint 13
Cisco Anyconnect Secure Mobility Client31, 151
Cisco VPN client 17, 52, 58
Crypto ACL 38, 39, 44, 48, 96, 100
crypto map33, 43, 44, 45, 48, 51, 56, 57, 58, 63, 64, 67, 69, 100, 101, 107, 108, 114, 115, 119, 125, 132, 133, 147, 148, 169, 170, 171, 173, 174, 177, 179, 180, 181, 183, 186, 187, 190, 192, 194, 224, 226, 229, 230, 232, 234, 236, 238, 241, 245, 247, 249, 252, 253, 255, 258, 260, 263, 265
Destination NAT 138
Diffie-Hellman Group 14, 99, 146, 147
DMVPN3, 4, 5, 8, 9, 26, 27, 28, 29, 34, 35, 36, 83, 84, 85, 86, 87, 88, 89, 90, 214, 215, 217, 219
Dynamic Crypto Map 48
dynamic outside IP address 34, 36
dynamic public IP .. 36, 37, 47, 59, 86, 107, 124
EIGRP11, 13, 21, 25, 28, 34, 35, 69, 71, 72, 73, 78, 79, 80, 87, 88
ESP 13, 42, 43, 100, 103, 150
Fortinet 13
GRE3, 4, 5, 9, 10, 11, 12, 13, 18, 19, 20, 21, 22, 23, 24, 27, 28, 34, 35, 64, 65, 66, 67, 68, 69, 70, 71, 72, 73, 74, 75, 77, 83, 84, 196, 197, 198, 199, 200, 201, 203, 205
GRE VPN3, 9, 18, 19, 20, 21, 22, 23, 34, 65, 69, 70, 71, 72, 75
Group Policy117, 118, 119, 124, 147, 148, 149, 154, 155, 156
Hash Algorithms 14
Hub-and-Spoke3, 4, 5, 6, 12, 14, 15, 16, 19, 22, 23, 24, 26, 33, 34, 35, 45, 49, 69, 73, 76, 83, 100, 104, 105, 108, 121, 126, 128, 176, 182, 200, 207, 228, 243

IKEv1 IPSEC 10, 12, 95, 142
ikev1 policy98, 99, 107, 115, 118, 124, 132, 146, 147, 224, 226, 230, 232, 234, 238, 241, 245, 247, 249, 253, 255, 258, 261, 263, 265
IKEv2 IPSEC ... 5, 10, 12, 31, 142, 145, 147, 148
Interesting Traffic14, 38, 39, 53, 60, 68, 69, 96, 100, 105, 108, 113, 122, 131, 136, 140, 141, 143
Internet Key Exchange 13
IPSEC3, 4, 5, 6, 8, 9, 10, 11, 12, 13, 14, 15, 16, 17, 18, 19, 20, 21, 22, 23, 25, 27, 28, 29, 33, 34, 36, 37, 45, 46, 47, 49, 51, 52, 54, 55, 56, 57, 58, 59, 60, 62, 64, 66, 67, 68, 69, 70, 73, 74, 75, 80, 81, 82, 85, 89, 90, 95, 96, 98, 99, 103, 104, 105, 108, 111, 113, 114, 115, 116, 118, 119, 121, 128, 129, 131, 132, 133, 134, 136, 137, 138, 142, 143, 151, 154, 161, 162, 164, 168, 172, 176, 182, 185, 188, 196, 200, 223, 228, 235, 240, 241, 245, 251, 257, 262
ISAKMP 13, 14, 40, 62, 97, 106, 123
isakmp group 54
isakmp policy40, 41, 50, 54, 61, 66, 67, 69, 73, 81, 89, 113, 169, 170, 172, 174, 177, 178, 180, 183, 186, 189, 192, 194, 196, 198, 201, 203, 205, 208, 210, 211, 215, 217, 219, 235
isakmp profile54, 56, 57, 60, 62, 63, 64, 81, 113, 114, 186, 189, 190, 192, 194, 208, 236
Juniper 13
keyring54, 55, 56, 59, 61, 62, 64, 81, 113, 114, 186, 189, 190, 192, 194, 208, 235, 236
Layer3 tunnel interface 10
LDAP 162, 163, 164
Microsoft Active Directory5, 53, 161, 162, 163
mirror access-list 38, 96
multicast11, 13, 20, 21, 23, 33, 34, 35, 36, 64, 86, 87, 215, 217, 219
NAT31, 33, 39, 40, 50, 53, 60, 69, 95, 96, 97, 105, 106, 108, 109, 110, 111, 113, 114, 117, 122, 123, 126, 127, 128, 131, 132, 134, 135, 136, 137, 138, 139, 140, 141, 142, 144, 153, 169, 171, 173, 175, 178, 179, 181, 184, 187, 191, 193, 195, 197, 199, 202, 204, 206, 209, 211, 212, 216, 218, 220, 236, 257, 258, 260, 262, 263
Next Hop Resolution Protocol 27, 83, 86

NHRP 27, 28, 83, 86, 87
NHS server 86
OSPF 11, 13, 21, 25, 34, 35, 71, 87
Palo Alto 13
Phase 1 IPSEC.14, 40, 42, 52, 54, 55, 56, 60,
   66, 73, 80, 81, 97, 106, 118, 123
Phase 2 IPSEC.14, 42, 43, 52, 57, 62, 66, 67,
   74, 82, 99, 100, 101, 107, 119, 125
pkts decrypt ..46, 102, 103, 121, 137, 141, 150
pkts encrypt ..46, 102, 103, 121, 137, 141, 150
Point to Point Tunneling Protocol 90
Policy NAT 138
Policy-Based IPSEC VPN 10, 11
Policy-Based VPN 3, 4, 12, 33, 37, 38, 64, 95
Policy-Based VPNs 9, 12, 68
PPPoE 33, 47
PPTP 4, 5, 90, 91, 92, 93, 94, 221
pre-shared key41, 48, 50, 55, 61, 69, 98, 99,
   107, 119, 124, 146, 147
QM_IDLE 45, 46
QoS 11
RADIUS 5, 164, 166
Remote Access VPN 12
Route-Based VPN 3, 4, 10, 11, 18, 20, 22, 64
RSA 5, 161, 164, 166, 167
RSA Server 161, 166
secure tunnel 16, 30
Security Association 14, 45, 98, 101, 150
shared secret keys 40
show crypto ipsec sa46, 102, 120, 137, 141,
   150
show crypto isakmp sa 45, 101, 149
site-to-site VPN8, 15, 54, 57, 71, 75, 111, 117,
   118, 121, 128, 133, 135, 138, 139, 142, 144,
   145, 146

Site-to-Site VPN3, 4, 5, 6, 12, 14, 15, 16, 37,
   46, 48, 49, 58, 64, 95, 105, 128, 134, 138,
   142, 168, 172, 188, 196, 223, 251, 257, 262
Sonic Wall 13
Split Horizon 87, 88, 89
split tunneling 17, 55, 116, 123, 126
Split-Tunnel Access Control List 116, 154
Spoke-to-Spoke VPN 26, 83
Static Crypto Map 48
static public IP address 36, 37
Static Route Tracking 129, 130
Static VTI 22, 35, 76
TACACS 5, 164
transform set 42, 43
Tunnel interface 18, 19, 23, 25, 67, 87
tunnel-group98, 99, 107, 115, 119, 124, 125,
   132, 133, 143, 148, 149, 156, 157, 164, 166,
   167, 225, 227, 230, 232, 234, 238, 242, 245,
   247, 249, 253, 255, 259, 261, 264, 266, 268,
   269
two-factor authentication 161, 164, 166
unicast IP traffic 13
Virtual Tunnel Interface..3, 4, 9, 22, 35, 65, 75,
   76
VPN Hairpinning 8, 108, 126
VPN software client 13
VPN tunnel10, 11, 13, 17, 18, 30, 32, 36, 37,
   38, 39, 47, 54, 55, 61, 63, 64, 84, 96, 103,
   117, 129, 134, 140, 142, 144, 151, 154, 155
VTI3, 9, 10, 11, 12, 13, 18, 20, 22, 23, 24, 25,
   28, 34, 35, 75, 76, 77, 78, 82, 83, 84, 208,
   210, 212
WebVPN 3, 29, 30, 31, 32, 151, 154

Printed in Great Britain
by Amazon